Alan McKirdy has written many popular books and book chapters on geology and related topics and has helped to promote the study of environmental geology in Scotland. Before his recent retirement he was Head of Knowledge and Information Management at Scottish Natural Heritage. He is now a freelance writer.

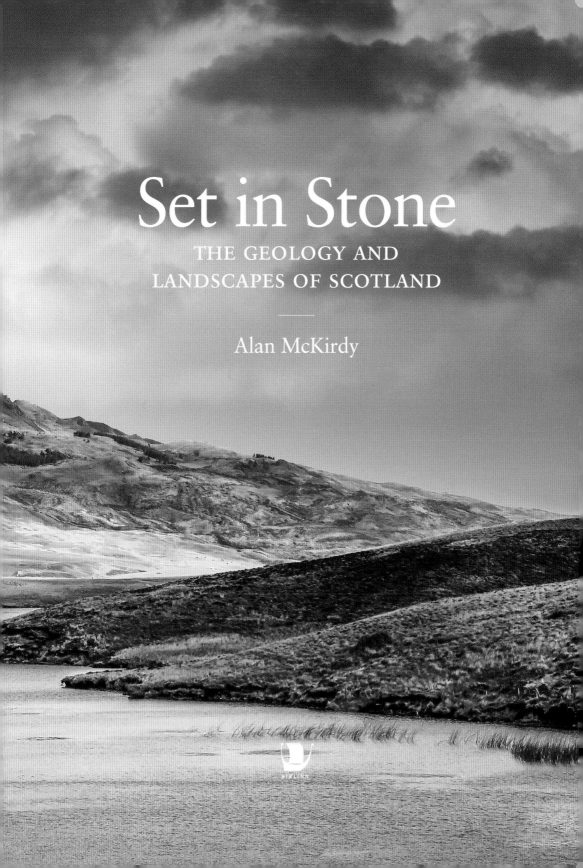

*For my three grown up children,
Stuart, Graeme and Fiona, of whom
I am extraordinarily proud.*

First published in
Great Britain in 2015 by
Birlinn Ltd
West Newington House
10 Newington Road
Edinburgh
EH9 1QS

www.birlinn.co.uk

ISBN: 978 1 78027 151 4

Copyright © Alan McKirdy 2015

The right of Alan McKirdy to be identified as
the author of this work has been asserted by him
in accordance with the Copyright, Designs and
Patents Act, 1988

All rights reserved. No part of this publication may
be reproduced, stored, or transmitted in any form, or
by any means, electronic, mechanical or photocopying,
recording or otherwise, without the express written
permission of the publisher.

British Library Cataloguing-in-Publication Data
A catalogue record for this book is available on
request from the British Library

Designed and typeset by Mark Blackadder

FRONTISPIECE.
The Old Man of Storr beyond Loch Leathan –
a landslip in lavas from the Skye volcano on the
Trotternish Peninsula, Isle of Skye

Printed and bound in Latvia by Livonia

Contents

Acknowledgements — vii

Unlocking the secrets of the stones – a personal journey — ix

Chapter 1 Scotland's journey in time and space — 1

Chapter 2 Scotland's story – from the beginning — 17

Chapter 3 Volcanic fires — 30

Chapter 4 Life on Earth — 48

Chapter 5 A geological legacy — 72

Glossary — 93

Further reading — 98

Index — 99

Picture credits — 102

Acknowledgements

I warmly acknowledge the help and support of my wife Moira who has been with me every step of the way and who also contributed her considerable biological expertise in commenting on all aspects of the book. Andrew Simmons and Mairi Sutherland from Birlinn were instrumental in transforming the text from a Word file to a finished book, and Hugh Andrew, for a second time, has shown faith in both the subject and the author. I have worked with book designer Mark Blackadder on a number of projects, and he always delivers an attractive fusion of words and pictures. A good friend for almost as long as I have been a geologist, Professor Stuart Monro OBE FRSE and more besides, made many useful comments and suggestions on the various drafts. I also acknowledge the comments made by Professor John Gordon in reviewing the draft text. John and I were colleagues from the late seventies onwards and we undertook many projects together, notably the precursor to this book – *Land of Mountain and Flood*. Professor Michael Usher OBE FRSE, formerly Chief Scientist at Scottish Natural Heritage, suggested additions to and deletions from the draft text with characteristic thoroughness and perception. My golfing buddy and languages expert Eric Gordon helped to sort out my punctuation and grammar. So I thank them all. But, as ever, any omissions or errors are mine and mine alone.

Opposite.
The Black Cuillin, Skye.

Unlocking the secrets of the stones – a personal journey

I am a geologist by accident rather than by design! I am one of those blessed individuals who was inspired by a very special school teacher – in my case, Mr Amos, my chemistry master. I filled in my course choice on my first day at university, with a degree in chemistry as the primary goal. Geology crept in under the wire as a subject that fitted in with that timetable. Initially geology and I weren't best friends. I had to wrestle with lots of unfamiliar new words that it seemed had been invented just to confuse and confound me personally. So, uncomprehendingly, I just learned the necessary facts and figures as if they were lines from a Shakespearean tragedy. The science fiction writer H.G. Wells, who briefly studied geology at university, found the subject to be 'a great array of damn cold assorted facts, lifelessly arranged and presented'. I would not be quite so rude about my *alma mater* and its teaching methods, but to begin with, it was a pretty similar experience for me. So, for those of you who are new to the subject, I understand your concerns. But stick with it; the pleasure and sense of discovery hugely outweigh the pain!

With the help of my second inspirational teacher Professor Nigel Trewin, I cracked the code and began to understand that the rocks reflected past life, environments and processes. So instead of learning that '500 metres of aeolian sandstone were to be found in the district of Elgin', I pictured a Sahara-like desert environment that existed around 250 million years ago with lumbering armour-plated reptiles making their way to a watering hole in the teeth of a withering desert storm. This was my eureka moment! Similarly, a coal seam became the remains of a verdant tropical rainforest of exotic trees and creepers, buzzing with dragonflies with wingspans of half a metre or more. Later, I studied the internal plumbing of the long-extinct Skye volcano and imagined a fire-spitting monster that, in its pomp, spewed lava and ash across the early landscapes. Much to my surprise, I

Opposite.
A typical landscape of the Northwest Highlands – a symphony of water and rock.

soon discovered that my home patch in the Borders had more than its fair share of extinct volcanoes too. It raised the question – if we had deserts and tropical rainforests in Scotland, then perhaps our climate hasn't always been characterised by horizontal rain, dreich days and watery sunshine. This opened the possibility, nay certainty, of past climates and environments being very different to our current experience. It quickly became all about seeing familiar landscapes and rock outcrops in a new and quite unexpected light. They all told a unique story, like the individual pages of the Earth's autobiography. After experiencing these revelations, any thoughts of a career as an industrial chemist were soon abandoned.

In writing this book, I would like to take you along a similar path to the one I trudged over forty years earlier. Push the jargon to one side and let's revel in the joy of the amazing story that Scots and those working in Scotland have helped to unravel over the last two hundred years. Moving continents, Scotland's journey across the globe, the concept of deep geological time are each an essential and, for the uninitiated particularly, a completely surprising part of the narrative. So suspend your disbelief and get stuck into one of the most amazing and largely untold stories that Scotland has to offer – the geological development of our country!

I hope you enjoy reading this book as much as I did writing it.

CHAPTER 1
Scotland's journey in time and space

In the following chapters, the amazing story of Scotland's geological evolution in time and space will be revealed. There are already many learned texts on this subject, but most assume a considerable prior knowledge of the subject. This book attempts to explain, in a way that everyone can understand and appreciate, the many wonders that the landscapes of Scotland have revealed during a century or more of study.

Piecing together the story from the stones is very much like detective work. Individual rock layers are pages in the Earth's history, and geologists from before the golden age of Victorian scientific inquiry to the present day have interpreted the evidence that the rocks provide. Each study builds on what went before. But it is also important to appreciate that what we currently hold to be an unassailable fact may well be revised and re-interpreted at some point in the future as new evidence comes to light.

Coastline of Orkney showing layers of sandstone that all tell a story in terms of the ancient environments they represent and the fossilised remains they may contain.

Erupting volcanoes were common in past times. Their eroded remains are prominent features of Scotland's landscape today.

And what a story there is to tell. The land that was to become Scotland experienced the absolute extremes that Nature has to offer. For example, the ancient crust of the Northwest Highlands and Hebrides was forged in the lower depths of the Earth's crust in conditions of extreme heat and pressure.

At other times in our geological past, oceans and lakes came and went like seasonal puddles. They left a build-up of sands, boulders and muds that hardened to strata, which later were fashioned by ice, wind and water into the familiar landscape features we recognise today.

Many of these rock layers contain evidence of past life that swam, burrowed, grew or simply walked on the Earth's surface – or Scotland's little corner of it. From the earliest and most primitive forms of life on Earth to an early version of a tropical rainforest that swathed the land around 340 million years ago, the range and complexity of life on Earth preserved in Scotland's strata is truly amazing.

Volcanoes were commonplace during many periods of our past history. Evidence of past volcanic activity is widespread throughout Scotland today, but, as we know, no volcanoes are currently active.

Evidence for the animals and plants that lived on the land that was to become Scotland has been preserved in the bedrock. They are revealed to us today as fossils. This panel illustrates a small part of the variety of life – or biodiversity – that existed through the geological ages. From top left to bottom right: a trilobite that inhabited the ocean deeps some 500 million years ago; a fossil fish of Devonian age that lived in a giant freshwater lake; fossilised plants from the Rhynie Chert; a crinoid or sea lily lived in the crystal-clear coastal waters; the very rare fossilised remains of a scorpion and a fossilised leaf from a tree that grew some 55 million years ago.

SCOTLAND'S JOURNEY

Another key aspect of the geological story of Scotland is the appreciation that, given sufficient time, continents can move great distances across the globe. The idea of plate tectonics has been around for fifty years or more and it is now accepted that, over extended periods of geological time, continents move around the globe roughly at the same rate as our fingernails grow.

As we observe a mountain range or familiar landmark in the countryside, it seems inconceivable that these monoliths are susceptible to change. They appear to have been there since time immemorial. But, with improved understanding of how the world works, we now recognise that the Earth's surface has always been on the move.

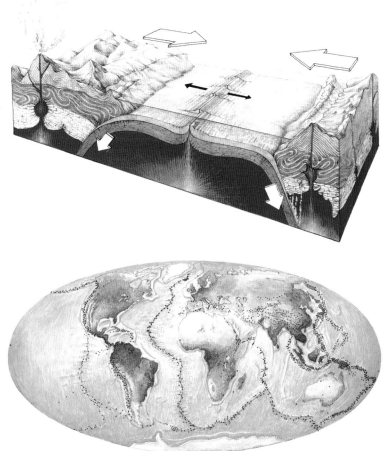

Right.
The oceans expand with the addition of new volcanic material at the mid-ocean ridges. But as the Earth never gets any bigger, there must be places where crust is destroyed. These are known as subduction zones, where the denser ocean floor dives under the continental plate. It is this constant interplay between creation and destruction of the Earth's tectonic plates that keeps the continents on the move – now and over the preceding billions of years.'

Right.
Earth's tectonic plates. Like a boiled egg that has been tapped sharply with a spoon, the Earth's crust is cracked into a series of segments. Geologists call these chunks of crust 'tectonic plates'. The plates are of uneven size, but together cover the full extent of the Earth's surface. Contact between the edges of the plates is inevitable, and earthquakes are created as a result of the constant motion. Some 1,000 earthquakes are recorded every day of the year.

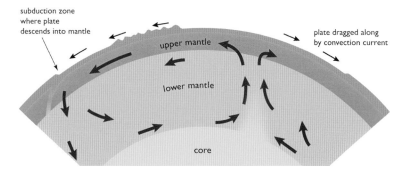

The Earth's surface consists of seven large rigid plates and about a dozen smaller ones. What drives the movement of the plates? For the answer, we need to visit the Earth's core. The constant ferment and movement within this molten ball of rock composed of the fairly common elements – predominantly iron and with some nickel – are what keeps our planet alive. The temperatures at the Earth's core are colossally hot, estimated to be around 6,000°C. As the molten metal churned and writhed in an attempt to cool down after the planet's formation when the Solar System was created, it set up movements in the overlying layer known as the mantle.

The mantle is a thick gooey layer that has the consistency of modelling clay and literally flows like a liquid in response to applied stresses. The heat radiating from the core sets up complex convection-type movements in the mantle that shunts the continents around.

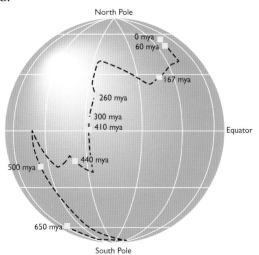

Above left.
Section through the Earth showing core, mantle and crust. The crust is the outer layer, shown in green.

Left.
Scotland's journey. From early beginnings close to the South Pole, Scotland travelled northwards to sit astride the Equator. With each change in global position, the climate affecting Scotland changed in response. Leaving the icy thrall of the South Pole, the climate warmed as the land approached the baking heat of equatorial latitudes. This journey continues to this day and into the future.

The individual plates are largely propelled by the convection currents created in the bowels of the Earth below. The plates bump and grind past each other in a series of jerky movements causing shoals of earthquakes, tsunamis and other natural disasters. Most of the earthquake action takes place at the margins of these plates. Most damage is done where a continent, carried as an integral part of one plate, collides with another landmass. Out of such collisions, mountains are born. The surface of the globe is littered with the results of such altercations – the Himalayas, for example, formed when India moved northwards and bumped into Asia. Closer to home, this is how the mountains of Scotland were created.

TIME

The formation of Scotland's geological foundations has taken place over an immensely long period of time. Scientists now believe the Earth is about 4,540,000,000 (4.54 billion) years old! Scotland boasts some of the oldest rocks to be found anywhere on the planet. Almost all of the periods between the most ancient to the most recent rocks are also represented in Scotland's geological record.

It's almost a freak of nature that such a small area of the planet's surface has such a long, rich and diverse geological history. This perhaps goes some way to explaining Scotland's hugely influential role in the development of the science of geology. Since the days of the Scottish Enlightenment in the late eighteenth century, when Dr James Hutton helped found the Royal Society of Edinburgh and wrote his seminal work *Theory of the Earth*, Scots and those working in Scotland have played a vital role in understanding the geological evolution of the planet.

Along with penicillin, steam engines, the telephone, TV, golf and whisky, Scotland also gave the science of geology to the world.

SCOTLAND'S 'TIME LORDS'

Understanding the immensity of geological time is something that scientists have grappled with over many centuries. Scots have made a particular contribution to this. The following pen-pictures describe those who made an outstanding contribution to the

James Hutton, the founder of modern geology

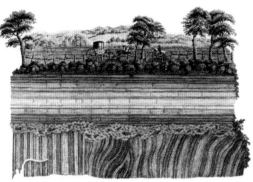

Left.
Siccar Point. The rocks in the foreground were tilted and buckled before the horizontal layers were deposited on top. Hutton understood that the processes required to create this feature would have taken aeons of geological time. It was from these simple observations that Hutton developed the concept of 'deep time'.

Right.
This illustration is of Allar's Mill near Jedburgh, drawn by Hutton's travelling companion John Clerk of Eldin. It illustrates the same relationships as at Siccar Point, where younger flat-lying strata overlie older up-ended rocks. This type of juxtaposition is known as an 'unconformity'.

unravelling of the mysteries of geological time. Let's call them Scotland's 'time lords'!

James Hutton is celebrated as our first 'time lord'. He was part of the Scottish Enlightenment – that golden age of Scottish history at the end of the eighteenth century when Scotland led the world in a multiplicity of disciplines – medicine, chemistry, economics and, of course, geology.

An early biographer said of Hutton 'It is no exaggeration to place Hutton beside Isaac Newton as being one of those rare scholars who opened an entirely new field to the human mind.' Hutton qualified as a medical doctor, but never pursued a career in this field. Instead he initially turned his hand to farming and also to the industrial production of sal ammoniac from soot, a raw material abundant in smoke-stained Auld Reekie. This chemical was used in the dyeing process and the working of tin and brass. But it was only when he reached the grand old age of 42 that Hutton set about seriously pursuing his real passion – a better understanding of the geological realm. By that time, he had set up home at St John's Hill in Edinburgh, which looks onto Arthur's Seat, a landscape now interpreted as the wreck of an ancient volcano.

For a period of three years from 1785, Hutton undertook a series of forays across the country to find evidence for his developing masterwork *Theory of the Earth*. His trips took him to Glen Tilt in Highland Perthshire, where he saw granite veins forming a network of off-shoots in rocks we now know to be older than the granite.

This observation demonstrated that granite wasn't the oldest

This plaque marks the spot where James Hutton's house once stood.

rock of all, which had previously been thought to be the case. The veining also proved that granite was at one time molten. This flew in the face of earlier 'wisdom' that all rocks were laid down at the bottom of an ancient ocean created during the biblical flood. An excursion in 1786 took Hutton and his field companion, James Clerk of Eldin, to Galloway, where the pair observed a similar phenomenon of granite veins from the Cairnsmore of Fleet granite running into the surrounding rocks. In Hutton's own words, 'we may now conclude that, without seeing granite actually in a fluid state, we have every demonstration . . . of granite having been forced to flow in a state of fusion'.

Hutton also visited the Isle of Arran, Jedburgh and most famously Siccar Point on the Berwickshire coast. His quest this time was to look for the point where rocks of different ages were juxtaposed. Today geologists call such features 'unconformities'.

The following description of his visit to Siccar Point was written by John Playfair, one of his companions on this visit. It is one of the classics of geological literature:

> On us who saw these phenomena for the first time, the impression made will not be easily forgotten. The palpable evidence presented to us, of the most extraordinary and important facts in the natural history of the earth, gave a reality and substance to our theoretical speculations. The mind seemed to grow giddy by looking so far into the abyss of time . . .

Although the interpretation of these natural features that Hutton observed in the field may seem narrow arcane debating points of little consequence, they remain some of the key foundation stones of the science of geology to this day.

His main conclusions were that:
- 'rocks from the volcanic fires', or igneous rocks as they were subsequently called, existed at one time in a liquid form;
- the action of water and wind can significantly change the surface of the Earth (what we now call 'erosion');
- these processes of erosion seen today operated in a similar manner in the geological past ('the present is the key to the past');
- the natural processes of the Earth acted to 'recycle' rocks (change their form from one rock type to another);
- and finally, and most importantly for Hutton's qualification as a 'time lord', the geological time over which these processes operated was incalculably long. He could see no vestige of a beginning and no prospect of an end to geological time. He could not put precise dates on the timescales he deduced from his observations because that particular technological advance had yet to be made. The world would need to wait another century and a half for absolute dates to be placed against specific geological events, courtesy of our third time lord!

Hutton marshalled his arguments in his two-volume work *Theory of the Earth with Proofs and Illustrations*, which was published in 1795. However, it was far from being a publishing sensation and its immediate impact on the scientific establishment fell considerably short of a barely detectable ripple. Hutton came across in conversation as a sparkling intellect and wit, but, on paper, he was dull and difficult to understand. It took the intervention of the next 'time lord', Sir Charles Lyell, for Hutton's ideas on how the Earth works to be heard and more widely understood.

Sir Charles Lyell was born in 1797, the same year that James Hutton died. Lyell was the first great science communicator. He made the many advances in geology available to everyone who cared to read his prolific output of textbooks. His most celebrated work was undoubtedly *Principles of Geology*. Although it wasn't the first book on geology, Lyell made the subject come alive and gave geology relevance beyond a narrow circle of enthusiasts. He took many of the ideas outlined by Hutton and gave them clearer meaning and significance. He was also well connected socially,

Hooker, Lyell and Darwin at Down House, Kent

which was important for scientists working during the early part of the nineteenth century. He counted Charles Darwin amongst his many friends and, along with Joseph Hooker, another prominent scientist of the day, was a frequent weekend visitor at Down House in Kent where Darwin lived for much of his adult life.

As Darwin travelled the globe on HMS *Beagle*, he took with him a copy of his great friend's book. 'I had brought with me the first volume of Lyell's *Principles of Geology*, which I have studied attentively; and this book was the highest service to me in many ways ... There are few authors who could write profound science and make a book readable.' It was of greater value to Darwin than that of just a good bedside read. Lyell's confirmation of the immensity of geological time allowed for the possibility of what was arguably the most important idea of the nineteenth century – Darwin's theory of evolution. Such a conjecture required many generations of a particular species to develop and evolve. Without the necessary idea of 'deep time', the concept of the origin of species by natural selection would have been dead in the water.

Lyell also struck a blow for the developing science of geology to take a secular view of the world. The final edition of *Principles* describes the Geological Society of London, near Piccadilly, as the place for lively debates about how the world came to be. Many geologists had, up until that point, ensured a consistency between Genesis, the first book of the Bible, and any geological theory they developed to explain the natural world. The biblical Great Flood was writ large in their published works. However, Lyell emerged as the principal advocate of the progressive scientists who sought to explain the physical world in terms of rational observations, rather than Scripture. There were many ranged against him in this debate, and these views still have many adherents to this day.

Lyell was born in Kinnordy House, near Forfar, but the family moved to Hampshire when he was still an infant. After graduating from Oxford with a degree in classics, he entered Lincoln's Inn to study law. As with Hutton, whose first career in medicine misfired, Lyell was later drawn more to the physical realm than legal entanglements. A problem with his eyesight contributed to this change of tack.

The first edition of *Principles of Geology* was published in 1830 and Lyell was appointed Professor of Geology at King's College in London a year later. He was knighted in 1848 and was finally laid to rest in the nave of Westminister Abbey in 1875 in recognition of his scientific achievements.

Arthur Holmes is our third 'time lord'. Although Scotland can't claim him as a Scot, he did much of his best work during his appointment as Regius Professor of Geology at Edinburgh University. Arthur Holmes was born in Gateshead in 1890 and went on to study physics at Imperial College in London. In fact, he never gained any formal qualification in geology!

His initial and enduring interest was in the age of the Earth and working out an absolute timescale for subsequent geological events. Much stuff and nonsense had already been written on this subject by the time Holmes had begun to take an interest. An Irish cleric, Archbishop Ussher, was the first to pronounce. After careful study of the Bible, he published an illuminated manuscript that proclaimed the Earth was formed on the 'twenty-third day of Octob. in the year 4004 BC'. This breathtakingly exact estimate, published in 1658, persisted as a valid figure for a couple of centuries despite the best efforts of Hutton and then Lyell to bring some scientific rigour to the debate. Around this time, Sir Isaac Newton used a much more scientific methodology to conclude that the Earth must have taken 50,000 years to cool from a molten state, but this finding was ignored as it was inconsistent with church teaching.

Next up was Lord Kelvin, a fiercely combative professor of natural philosophy at Glasgow University. He dominated the world of physics for a good part of the late nineteenth century. In fact, it was an exchange in *The Times* newspaper between Kelvin and the next generation of researchers about the age of the Earth that caught the young Arthur Holmes' attention whilst he attended Gateshead Higher Grade School. Although Lord Kelvin was eighty-two by this stage, he still clung valiantly to his calculation that the Earth was no older than 20 million years. The new work reported in *The Times* relied on an understanding of radioactivity. This was to hold the key to the new methodology of accurately calculating the age of our planet as well as individual rock specimens. The science of geochronology – the accurate dating of rocks and past events – was born.

Radioactivity in rocks is an entirely natural phenomenon. It may sound like medieval alchemy, but some elements are inherently unstable and change from one form to another through radioactive decay. Uranium, for example, decays over a known 'half life' period when half the mass of a given sample decays from uranium to a decay product – in this case lead. The half-life of some elements is short and can be measured in terms of a few seconds, whilst other elements take many millions of years to lose half their original mass. Uranium has a very slow rate of decay, taking four and a half thousand million (or billion) years to lose half of its mass.

Scientists, from Marie and Pierre Curie onwards, have established which elements decay to what products and the period over which this transition takes place. So we measure the amount of original material (e.g. uranium) remaining in a sample and how much decay product (e.g. lead) is detected in the same sample. Knowing the half-life over which the decay takes place, we can estimate how long this process has been going on. This number is then taken to be the time when a particular rock sample was formed, or in the case of igneous rocks (rocks that owe their origin to volcanoes), the point at which the rock cooled to a specific temperature, known as the Curie Point.

This lab-based dating process revolutionised the science of geology because, for the first time, we could put an absolute number on past events. Up to this point, dating was relative: we could say this rock here is younger or the same age as that one 10 kilometres away by looking at its fossil content, relative position and by gut feeling, but we couldn't say by how much. Radioactive dating changed all that. It also shone a light on the age of individual rock samples and, most importantly, on the absolute age of the Earth. So the clock had been found; all that scientists needed to do now was to learn to tell the time! This was easier said than done.

Holmes' first major success in the dating game was to analyse a lump of rock from Norway using the uranium–lead method. It was the first time such a determination had ever been made. He

Sir Charles Lyell

Professor Arthur Holmes

The new 'Time Lords' gallery at Our Dynamic Earth in Edinburgh features a 'conversation' between these three venerable gentlemen of science. They discourse on the difficulties of understanding the immensity of geological time and bring to life the contribution they each made to this important debate. Dr James Hutton in his Edinburgh study – a scene admirably recreated by Professor Stuart Monro, formerly Scientific Director at Our Dynamic Earth.

published an age of 370 million years for this rock. He followed that up with a series of age determinations on rocks from the older age of the spectrum – Carboniferous, Devonian, Silurian and Precambrian. His results still fit comfortably with our current estimates for the ages of the various component parts of the geological column (see the next section), which were derived using much more sophisticated analytical equipment and techniques. This was a staggering achievement for this truly remarkable pioneer of this new area of science.

Arthur Holmes is also celebrated for his many other achievements including an early model of plate tectonics and the publication of a textbook entitled *Principles of Physical Geology*, which ran to many editions. Every serious student of the subject has a copy on their shelves. But Holmes will be forever associated with 'The Dating Game', which is incidentally the title of Cherry Lewis' excellent biography of this exceptional man.

As a fitting tailpiece to his life and considerable achievements, Holmes said 'Looking back, it is a slight consolation for the disabilities of growing old to notice the Earth has grown older much more rapidly than I have – from about six or seven thousand years when I was ten, to four or five billion years by the time I reached sixty!'

SCOTLAND THROUGH THE AGES

These geological periods are referred to in the chapters that follow, so you may find it helpful to mark this page and refer to it as necessary.

Period of geological time	Millions of years ago	Scotland's position on the globe	Environment and events
Anthropocene (broadly equivalent to the Holocene)	Last 10,000 years	57° N	This is the 'Age of Man' where much of humanity's history is recorded. *Homo sapiens* (us!) has changed the world about us in a fundamental way by, for example, modifying ecosystems and exploiting the planet's natural resources on land and in the oceans. This has been true particularly since the Industrial Revolution 250 years ago.
Quaternary	Started 2 million years ago	Present position of 57° N	This was the period when the landscape we are familiar with today was sculpted by ice. This episode is popularly known as the Ice Age. Scotland sat at the edge of the polar ice cap that extended from the North Pole. It wasn't just one deep freeze; there were many advances and retreats of the ice, punctuated by warmer times called interglacials. We are in an interglacial at the present time. But in due course (currently estimated at around 50,000 years hence), the thick cover of ice and snow will return and all living things will need to adapt to the new conditions – or perish.
Neogene	2–24	55° N	After the tropical heat of the previous period, the country started to descend into the deep freeze of the Ice Age.
Palaeogene	24–65	50° N	As the North Atlantic Ocean continued to widen, the Earth's crust thinned and a chain of volcanoes burst through to the surface. St Kilda, Skye, Mull, Ardnamurchan, Arran and Ailsa Craig were all fire-spitting volcanoes active at this time.

Period of geological time	Millions of years ago	Scotland's position on the globe	Environment and events
Cretaceous	65–142	40° N	This was the age of the chalk in England, although few deposits of this age are to be found in Scotland now. A lump of chalk the size of a semi-detached house has been found on Arran, suggesting that there could have been a covering of chalk deposits across parts of the country that was later eroded by ice, wind and water.
Jurassic	142–205	35° N	This was the age of the dinosaurs. Both flesh- and plant-eating dinosaurs roamed the land. Their remains are particularly abundant in the sediments found on Skye.
Triassic	205–248	30° N	During both of these periods, Scotland sat at a similar position to the Sahara desert of today. So great sand dunes were common with occasional oases where early reptiles gathered to drink.
Permian	248–290	20° N	
Carboniferous	290–354	On the Equator	Scotland sat astride the Equator during this period. Verdant tropical rainforests flourished that provided a home to many exotic species of plants, reptiles and insects.
Devonian	354–417	10° S	The land that was to become Scotland sat in the middle of a large arid continent. Desert conditions prevailed as many layers of sandstones (the Old Red Sandstones) accumulated on land and at the bottom of a long-disappeared lake. The Glencoe super-volcano breathed fire during these times.
Silurian	417–443	15° S	The final stages of the closure of the Iapetus Ocean took place creating a united landmass.
Ordovician	443–495	20° S	Many layers of mud and sand accumulated on the floor of the Iapetus Ocean. During this time, the ocean began to close and continents collided with an almighty crunch, leading to the formation of the Highlands.
Cambrian	495–545	30° S	Life exploded into a variety of forms during this time. Durness Limestone dates from this period.

Period of geological time	Millions of years ago	Scotland's position on the globe	Environment and events
Proterozoic	545 – 2,500	Close to South Pole	This chunk of geological time spans the formation of much of the land that now forms the uplands of Scotland. The Iapetus Ocean opened during this time and many layers of sedimentary rocks accumulated within its waters. The ocean floor was partly composed of Lewisian gneiss. The earliest forms of life are recorded in rocks from this age. Torridonian sandstones and Moine schists were formed during this time.
Archaen	Prior to 2,500	Unknown	Formation of the Lewisian gneiss deep within the Earth crust. Evidence of volcanic activity is also recorded.

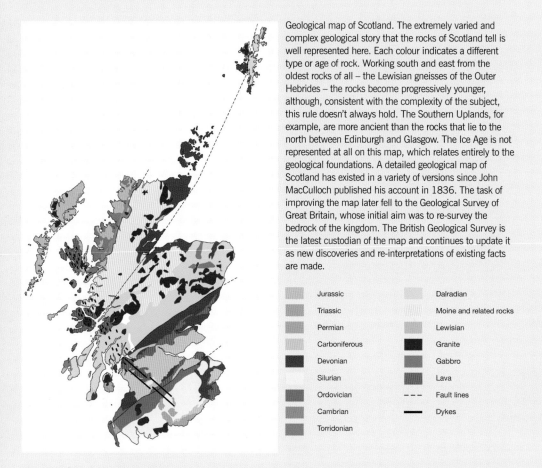

Geological map of Scotland. The extremely varied and complex geological story that the rocks of Scotland tell is well represented here. Each colour indicates a different type or age of rock. Working south and east from the oldest rocks of all – the Lewisian gneisses of the Outer Hebrides – the rocks become progressively younger, although, consistent with the complexity of the subject, this rule doesn't always hold. The Southern Uplands, for example, are more ancient than the rocks that lie to the north between Edinburgh and Glasgow. The Ice Age is not represented at all on this map, which relates entirely to the geological foundations. A detailed geological map of Scotland has existed in a variety of versions since John MacCulloch published his account in 1836. The task of improving the map later fell to the Geological Survey of Great Britain, whose initial aim was to re-survey the bedrock of the kingdom. The British Geological Survey is the latest custodian of the map and continues to update it as new discoveries and re-interpretations of existing facts are made.

- Jurassic
- Triassic
- Permian
- Carboniferous
- Devonian
- Silurian
- Ordovician
- Cambrian
- Torridonian
- Dalradian
- Moine and related rocks
- Lewisian
- Granite
- Gabbro
- Lava
- --- Fault lines
- — Dykes

CHAPTER 2

Scotland's story – from the beginning

The Great Tapestry of Scotland celebrates all that is best about the country. So it is fitting that the story of the tapestry starts with 'The Formation of Scotland' and 'The Collision'. It is every earth science communicator's dream that the depiction of Scottish history and heritage doesn't start with the Picts or at Skara Brae in Orkney. It usually does! But the story of the collision of Scotland with England and the associated crumpled rocks of the

Panel 3a of the tapestry shows rock slashed into segments running north-east to southwest, which is a fitting representation of the bedrock of Scotland. This is the geological fabric of the land. Some of the key geological gems are picked out: the Lewisian gneiss – the basement on which all younger rocks are perched; the Torridonian sandstone peaks of Slioch and Suilven; the mangled Dalradian rocks, which were folded as the Iapetus Ocean closed 420 million years ago to unite these islands forever.

Panel 3b depicts the peaks of North Berwick Law, Arthur's Seat, St Kilda and those on Skye, which all breathed fire in the past. We also see the Permian deserts of Mauchline and Dumfries represented and the stumps of Lepidodrendron trees, a proxy for the tropical rainforests that once cloaked the land where Edinburgh and Glasgow now sit. Some of the key processes that gave rise to this rich geological heritage are also depicted – the eroding coastline and sea-stacks and the crashing of the waves. It has echoes of one of James Hutton's biggest ideas – that the 'present is the key to the past'.

Highlands still requires a degree of explanation for those who may not yet have heard the tale. That is the purpose of this chapter.

FROM ANCIENT TIMES

In starting Scotland's great geological odyssey, we must rewind almost to the beginning of geological time. Our planet formed 4.54 billion years ago, but the oldest rocks in Scotland, the Lewisian gneisses, date back a mere 3 billion! These bent and buckled rocks were forged in the Hadian fires of the lower crust. Through a combination of earth movements and erosion, the rocks are now visible at the surface. The Outer Hebrides and Northwest Highlands are the best places to see these most ancient vestiges of our geological past.

The Outer Hebrides are a chain of islands around 200 kilometres long, stretching from the Butt of Lewis in the north to Barra in the south. They have a rare, untainted and rugged beauty. Nature rules here.

The indomitable forces of erosion have gnawed away at the land to the point where the bare bones of the rock beneath are left raw. This unceasing wearing down has left a landscape bereft of mountains of any great consequence. What remains is still dramatic – a sweeping vista of smooth rocky knolls studded with

The landscape has been scoured and scarred by the passage of ice in recent times. But the bedrock is ancient – one of the oldest pieces of the Earth's crust to be found anywhere on the globe. Sea-level rise since the ice melted has partially flooded this part of the Western Isles.

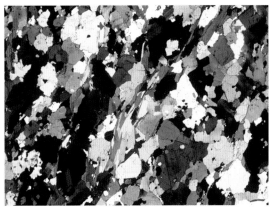 

Left.
Thin section or slice through Lewisian gneiss showing the different component layers under the microscope. Geologists routinely take these type of slices to determine the mineral composition of rocks.

Right.
Lewisian gneiss showing the characteristic bands of light and dark minerals. This banding was created by intense heat and pressure when the rocks were buried deep in the Earth's crust.

irregularly shaped deep pools of dark, peaty water.

In the northern part of the archipelago, and in the southern isles – the Uists – this 'knock and lochan' landscape is lost, buried beneath a thick blanket of peat that is just the most recent addition to a truly remarkable history that stretches back billions of years.

The Outer Hebrides are strung out along a fragment of the Earth's crust that is of unimaginable antiquity. These islands date back to one of the earliest phases of our planet's existence, when the primordial 'Fires of Hell' were still 'burning'. With an immature, unstable crust and a fledgling atmosphere unsuitable for most forms of life, the early Earth was an inhospitable place.

The rocks we see at the Hebridean surface today were formed many kilometres below, destined only to see daylight after the upper layers were planed away and exhumed by eons of the erosive forces of wind, water and ice. At such depths, the pressures were enormous, separating out dark and light minerals into discrete bands.

The light-coloured bands are made up largely of the commonest rock-forming minerals – white quartz and feldspar – with the darker bands containing metamorphic make-overs of the iron- and magnesium-rich minerals, which include the dark-green mineral amphibole. The resulting panchromatic banding gives us one of the highest temperature and pressure grades of metamorphic rock that exist: gneiss (pronounced *nice*).

The fabulous swirls and intricate contortions of the Lewisian gneiss stand testament to a long and tortuous history. Amidst such disruptions comes the injection of molten rock, sheets of magma

Top.
This road cutting north of Ullapool is one of the best places to see the various component parts of the Lewisian gneiss. Much of the bedrock here is known as 'grey gneiss' which gives a clue as to its appearance, whilst the darker bands represent pulses of molten rock injected at a later date than the formation of the grey gneiss. The final act was the injection of pink granite that cuts across both the grey gneiss and the darker bands.

Bottom.
The west coast of Harris is hewn from granite. Over millennia, even this toughest of rock has been battered and cut into jagged sea-stacks.

cutting through the layers of gneiss and solidifying in criss-crossing ribbons, known as 'dykes'. The fact that dark dykes slice across the grey gneiss gives us the clue that the igneous intruders are younger, and it is from such observations that a relative sequence or chronology of events builds up and allows us to reconstruct the remote history of these times.

However, the Western Isles are not all gneiss. Other rocks have left their legacy here. The upland areas of South Harris and western Lewis rise on the back of great sheets of pink granite. This magnificently tough granite forms towering cliffs and jagged sea-stacks off the west coast of Harris, home to a myriad of small uninhabited islands, known as skerries.

Roineabhal – let's leave well alone

In the mountainous southeastern corner of Harris is a rock that has more in common with the Moon than with our Earth, for the lunar uplands are made of the same feldspar-dominated anorthosite bedrock that is found here.

The rock from which Roineabhal on Harris is made is ideal for the purposes of road-building and other uses in the construction industry. In the early 1990s the area was not surprisingly targeted as a potential site for Scotland's second superquarry, to help feed the voracious ever-growing international appetite for crushed rock aggregate. But the application was unsuccessful, and the natural beauty of Roineabhal remains untarnished. Indeed, much of this terrain has been designated as a National Scenic Area because of its outstanding landscape value. It is almost treeless, its very barrenness being part of its attraction. The elements of sea, freshwater, rock and moor are intimately woven together into a tableau of stunning natural beauty.

Roineabhal, South Harris – proposed site of a superquarry in the 1990s. The project didn't proceed at that time, but it remains a tempting prospect for the minerals industry.

Although Lewis in the Outer Hebrides is the area after which the Lewisian gneiss is named, gneiss is also found elsewhere in Scotland – mainly in the Northwest Highlands.

RIVERS OF SAND

The next dramatic episode in the geological story is that this ancient landscape of the Lewisian gneiss became lost from view for many millions of years almost in the geological instant. Great streams and rivers, carrying vast quantities of sand, boulders and other debris eroded from the mountains, flowed to the sea across the desolate early landscape. Over a relatively short period of geological time, a blanket of debris over 5 kilometres thick was dumped – now hardened to sandstone and conglomerate. This is the Torridonian sandstones, which formed at a time when Planet Earth was largely devoid of life of any sort – plants or animals.

Erosion has cut through this cover of sandstone in many places, and, in a few, it has reached the precise level of the junction between the Torridonian and earlier Lewisian crust below. In these places, we can see, touch and walk over a Lewisian land surface that existed over a billion years ago and we can experience its original valleys and knolls, as they were before the Torridonian rivers came flooding in. The lower slopes of the towering peak of Slioch, at the southern end of Loch Maree, provide just such an opportunity, and the northern shore of Upper Loch Torridon is another location where this ancient land surface has been exhumed by much more recent erosion by ice, water and wind.

Detailed study of these rocks reveals other aspects of this long-lost world. The climate on this slice of Planet Earth one billion years ago can be deduced with a degree of confidence. The identification of tropically weathered soil horizons and other environmental indicators suggests an equatorial climate, with hot dry summers during early Torridonian times.

ROCKS OF THE MOOR – THE MOINES

The Northern Highlands, bounded to the south by the Great Glen and to the west by a significant break in the Earth's crust known as the Moine Thrust, are largely built from a thick sequence of altered rocks known as the Moines. The name is derived from

Left.
Slioch, on the north side of Loch Maree, helps us picture an ancient landscape that existed a billion years ago. The lower slopes of this mountain are comprised of Lewisian gneiss. They gently undulate, creating a wide valley. Then came a deluge of sandstone laid down during Torridonian times, burying this Lewisian landscape. It was only in recent times that the cover of sandstone was substantially reduced by ice erosion and the Lewisian bedrock was once again laid bare at the surface.

Right.
Even amongst the scenic grandeur of the peerless Northwest, Suilven stands out as an aristocrat amongst Scottish mountains.

Present-day landscape of the Moine – windswept and boggy.

the Gaelic *móin*, meaning moor or bog, graphically describing the type of habitat the area now supports.

Many layers of sands and muds were laid down in a long-disappeared ocean over many millions of years. These strata are of broadly similar age to the Torridonian sandstones. However, unlike the sandstones, the Moines were subsequently cooked and squashed by deep burial in the Earth's crust. This took place around 800 million years ago during mountain building. These events produced a series of changed or metamorphic rocks of great durability.

The Moines are folded, buckled and shot through with dazzling patches of white quartz and pink-flecked seams of granite. Puncturing this blanket coverage of Moines is a series of much larger masses of granite and related intrusions, squeezed upwards during the Caledonian mountain-building period, when the Highlands of Scotland were born. Intense heat and pressure generated in the lower crust caused local melting, and great balloon-shaped masses of molten granite magma, less dense than the surrounding rocks, ascended the crust. Erosion immediately after the mountains were formed, and over the millennia since then, has unroofed these once molten masses to reveal coarsely crystalline granites of differing compositions. Each has its own signature composition of minerals that determines the rock's colour and general appearance.

THE IAPETUS OCEAN – SCOTLAND'S CRUCIBLE

It's perhaps a difficult concept to grasp, but in these far off times, much of the land that was to become Scotland didn't exist! How the bedrock of the Highlands and southern Scotland came to be is the next part of the story. We know from centuries of study that the most ancient rocks that form much of the Highlands of Scotland and the Southern Uplands were originally laid down as sediment, primarily sands and mud, in an ancient and long-disappeared ocean. It was called the Iapetus Ocean, after Titus Iapetus, father of Atlas in Greek mythology.

The Iapetus Ocean separated two halves of what is now the United Kingdom – Scotland and Northern Ireland from England and Wales. The ancient basement of the Lewisian, Torridonian and Moines, which existed at this time, formed a small part of

The Southern Uplands are largely built from the sediments that were swept into the Iapetus Ocean and later lightly cooked and squashed as the ocean closed and continents collided. The landscape has none of the extravagant peaks of the Grampian Mountains to the north, as the rocks of the Southern Uplands are softer, so more heavily moulded by the passage of ice.

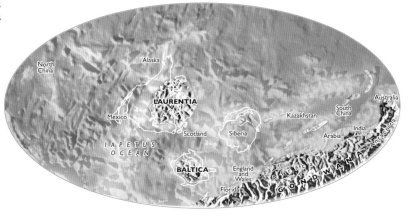

The geography of the world 600 million years ago looks very unfamiliar. The Iapetus Ocean separated Laurentia, which included proto-Scotland, from Baltica (Norway, Sweden and Finland) and Avalonia, which included England and Wales. Avalonia was a small part of Gondwana, a continental assortment that included South America, Africa, Arabia, India, Australia and South China. The Iapetus Ocean existed for around 200 million years.

the North American continent Laurentia, whilst what was to become England was marginal to the larger landmass of Gondwana. This view of the world requires a great stretch of imagination, but there is good evidence to support the reconstructed geography of the early planet as illustrated here. It seems more plausible when you consider that the continents are constantly on the move and have been since the Earth's crust formed.

It is also logical to assume that the distribution of land and sea as represented in today's world atlas will change out of all recognition in millions of years to come. In studying the history of the planet, the golden rule is that change has been a constant.

THE OCEAN CLOSED

The Iapetus Ocean reached its greatest extent around 480 million years ago when it was around 5,000km wide. Thereafter, it began to close as the continents were once again on the move. This process happened in stages. Firstly Laurentia, which encompassed modern-day North America and Scotland, collided with Avalonia, the precursor of England and Wales, and finally with Baltica. This continental car-crash created a mountain pile-up of Himalayan proportions that stretched from north Norway to the

The Caledonian mountain belt. These mountains were created by the collision of continents. The Caledonian Mountains stretch from the far north of Scandinavia though the British Isles, on to Newfoundland and terminate as the Appalachians in America. The opening of the Atlantic Ocean many millions of years after the formation of this mountain belt split it asunder, so that its remnants now appear on opposite sides of the world.

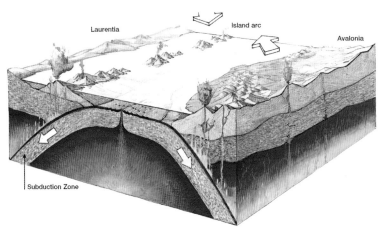

The Iapetus Ocean began to close in early Ordovician times, some 480 million years ago. A chain of volcanic islands (an island arc) developed offshore from 'Scotland' as the descending oceanic plate melted and returned to the surface as molten lava. It was the island arc that first collided with Laurentia, squashing the sediments that had been deposited in the Iapetus Ocean. This vice-like grip altered these sands and muds into a towering mountain chain, the eroded remnants of which we now recognise as the Highlands of Scotland.

Appalachians in the south. Today we can recognise a sliver of this once mighty mountain chain as the Highlands of Scotland.

As the ocean closed, the pile of muds and sands that had accumulated in the ocean deep were sliced into wedges and dragged into the abyss as the continents converged. As the plate on which the sediments accumulated dived deeper into the Earth's crust, the overlying rocks melted, creating an arc of volcanoes at the surface. It was the collision of this island arc with Laurentia that caused much of the damage. This crashing together of two irresistible objects created the conditions under which the sediments were changed from sands and muds to the towering peaks of the Caledonian Mountains. These now highly changed or metamorphosed rocks are known as the Dalradian, named after the old Celtic region of Dal Riata (Dalriada). Their extent stretches from the Great Glen Fault in the north to the Highland Boundary Fault in the south.

The volcanoes of the island arc are now long gone, removed from the geological record by erosion, but evidence of their

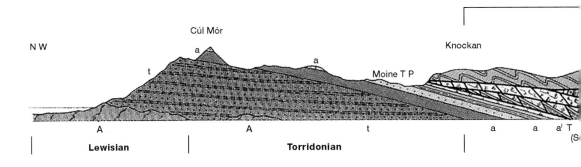

Pettico Wick near St Abbs Head shows folds and buckles in the rocks that resulted from the continental collision between Scotland and England. The layers of mud and sand laid down on the ocean floor of the Iapetus Ocean were contorted as the continental plates came together.

presence is preserved as pebbles in the accumulated sands and muds of the Southern Uplands. Also layers of ash, which would have been ejected from the chain of volcanoes, fallen into the adjacent waters of the ocean and accumulated on the ocean floor, are also identifiable at some locations.

The fabric of the Southern Uplands was also created at this time. The layers of sand and mud that built up on the floor of the Iapetus Ocean were folded and buckled as the continents collided.

THE FINAL ACT

The mark made by the final collision with Baltica is most clearly recorded in the Northwest Highlands. As the ancient continents ploughed into each other, powered by movements within the Earth's interior, great masses of rock rode roughshod over one another. The crust was concertinaed and a suite of great, almost

This diagram accompanied the memoir of the Geological Survey prepared by the celebrated Victorian geologists Ben Peach and John Horne. The concept of plate tectonics had yet to be developed, so the idea that slabs of older rocks could be driven for many miles along low angle faults in this manner was ground-breaking.

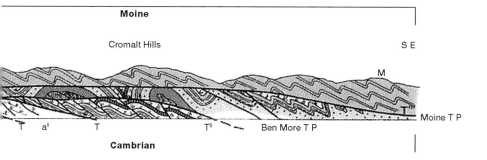

The north shore of Loch Glencoul provided Peach and Horne with strong evidence that great slabs of rock had indeed been driven westwards, as depicted in their cross section figured on the previous page. Above the dotted line, drawn to coincide with the Glencoul Thrust, a great slab of Lewisian gneiss has been drive westwards as continents collided and the Iapetus Ocean finally closed. This view is one of the best places to appreciate these massive earth movements on a landscape scale.

horizontal breakages developed, known as thrust faults, as vast masses of rock were bulldozed over each other. On the mainland, the famous dislocation that formed at this time was the Moine Thrust. It runs from the north coast of the mainland, near Loch Eriboll, to the Point of Sleat on Skye.

KNOCKAN CRAG

The rocks of the Northwest Highlands are amongst the most intensively studied anywhere in the world. Historically, a number of key geological phenomena were first observed and correctly interpreted in the Northwest Highlands and this helped to develop the science of geology internationally. It is also the place where many of our early geological pioneers cut their teeth.

Giants of the Victorian scientific world, such as Benjamin Peach and John Horne, were sent to this furthest corner to provide a definitive interpretation of the ancient and complex rocks of the Assynt area. This they did in their seminal memoir

Ben Peach and John Horne were stars of the Geological Survey. With a few colleagues, they spent months in the field each year slowly unravelling the mysteries of Scotland's geology. They are pictured at Inchnadamph Hotel in the Northwest Highlands after a hard day walking the surrounding hills and glens.

Scottish Natural Heritage has a vital mission to 'foster an understanding of the natural heritage of Scotland'. The visitor centre at Knockan Crag, north of Ullapool, tells Peach and Horne's story and also interprets the geology of the area.

published in 1907 by the Geological Survey. These were halcyon days, when the rest of the world looked to British pioneers for a lead in developing the nascent science of geology.

Scottish Natural Heritage's visitor centre at Knockan Crag, where the Moine Thrust is most famously exposed, celebrates the great achievements of Peach and Horne in unravelling its mysteries. The intricacies of the Knockan story had previously been just for the cognoscenti to appreciate. But the centre's narrative is told in such a way that everyone can understand the nature of the debate that took place between the early pioneering geologists when they wrestled with competing theories to explain the geology at this place. There is also a car trail called *The Rock Route*, which introduces a series of twelve locations that illuminate different aspects of the Moine Thrust story.

NEW WORLDS

After the dramatic convulsions involved in the closing of the great Iapetus Ocean, the remainder of our geological history played out at a rather more sedate pace. Scotland drifted ever northwards, driven by the vagaries of the Earth's shifting tectonic plates. As it moved northwards from its location close to the South Pole, the land that was to become Scotland passed through every climatic zone. So to see the remnants of desert sands in our record of the rocks or coals that mark the place where tropical forests once flourished should come as no great surprise. The plants and animals that inhabited this new world changed in response to Scotland's position on the globe. How the ancient ecosystems adapted to this journey will be explored in more detail in a later chapter.

CHAPTER 3
Volcanic fires

Arthur's Seat is an ancient and long extinct volcano located right in the heart of Edinburgh.

Most of those who have experienced the exhilaration of scaling the peak of an active volcano such as Mount Etna or Vesuvius will return home from these far-flung places grateful that the hills near where they live are less prone to such extreme behaviour. But many familiar Scottish landmarks hide a fiery secret – they too were once active volcanoes. Here are a few well-known landmarks with a volcanic past – St Kilda, Ben Loyal, Glen Coe,

the Skye Cuillin, Ailsa Craig, Goat Fell on Arran, the Campsie Fells, the Ochil Hills, the Sidlaw Hills, Arthur's Seat, North Berwick Law, Traprain Law, the Eldon Hills and the Cheviot Hills. This is but a small selection of a much longer list of familiar places that owe their origin to long-extinct volcanic activity.

This is the joy of studying how the landscape formed – seeing a favourite vista, such as the tranquil view over Arthur's Seat, and imagining the scene when things were at their most explosive! It would have been like any contemporary volcano – long periods where absolutely nothing happened, punctuated by periods of intensive activity when lavas were coughed up out of vents, holes or fissures in the ground and clouds of ash, stream and fusillades of boulders and volcanic bombs were hurled high into the sky.

What we see today in the ancient volcanoes of Scotland is not the classic cone shape of Etna. Our view is more likely to be that of the internal plumbing of the volcano – the magma chamber

Scott's View, reputed to be one of Sir Walter Scott's favourite vistas, has the Eildon Hills centre stage and the winding River Tweed in the foreground.

where the hot rocks accumulated before eruption, the offshoots of molten rock that were injected into the adjacent sediments or the associated lava flows. The action of the ice from a much later time tore lumps from the ancient structures and excavated deep into their heart. The Eildon Hills, towering over Melrose in the Scottish Borders, are perhaps the closest to having the classic cone shape of contemporary volcanoes, but that shape is entirely the work of ice moulding the solidified pulses of magma into the signature volcano profile.

What we are left with now is an incomplete story, but it's not too taxing to piece it back together and establish which elevation of the original structure the erosive forces of ice, wind and water have left behind for us to study. Across the country, it's quite a varied catalogue of aspects of the original volcanoes that we can now see.

Here are a few examples (see photographs opposite):

- Edinburgh Castle (top left) sits atop the volcanic pipe that connected the surface to the magma reservoir beneath.
- Arthur's Seat (top right) where a number of key elements of the ancient volcano are still visible – the volcanic vent choked with ash and volcanic bombs that didn't quite make it to the surface, the associated lava flows best seen on the north flank of Whinny Hill and Salisbury Crags, a later off-shoot of molten rock, now seen as sandwiched between layers of sedimentary rocks.
- Traprain Law (middle left) is another blister of molten rock (magma) that didn't quite make it to the surface. A few metres more and this reservoir of molten rock would have burst forth and flowed across the land surface.
- The Black Cuillin of Skye (middle right) is the deep-seated pool of magma that collected beneath the superstructure of the volcano and the Trotternish Peninsula to the north is partly comprised of layer upon layer of lavas that were erupted from the same volcano
- Between Auchterarder and Crieff, a dyke (upstanding sliver of igneous rock) was first noted by James Hutton on his travels across Scotland (bottom). Amazingly, it is now thought to be an offshoot from the Mull volcano some 130km to the west!

Following James Hutton's lead that 'the present is the key to the

past', scientists have studied – both under the microscope and in the great outdoors – the many active volcanoes and the rocks they produce.

A huge variety of lava types have been identified – from the black glassy obsidian to the commonly occurring basalt, rich in calcium and magnesium. All of these rocks from the present day can be found in the geological record. By studying active volcanoes today, we can learn most of what we need to know about the geological past, such as the complexity of these ancient volcanoes and the changing composition of the magma they produced.

Geologists use thin slices of rock, examined under the microscope, to identify component minerals and other attributes. Rocks, which may look dull in fist-sized specimens, appear as a kaleidoscope of brightly coloured minerals under the microscope. The mineral composition and structures seen under the microscope allow rocks to be classified and their origins to be determined. 'Thin sections' are therefore a key tool in the geologists' armoury.

The rocks illustrated here are
top left: gneiss,
top right: andesite,
middle left: gabbro,
middle right: mica schist
and bottom: a variety of limestone.

Rocks derived from processes related to volcanoes are known as igneous.

It was another Scot who pioneered the study of igneous rocks. Sir Archibald Geikie began a career in banking, but changed direction to join the Geological Survey. His interests in the field of geology were many and varied. They included a treatise on the glacial drift of Scotland, which was the first to recognise the connection between the work of ice and the way the country looks today.

Geikie's book *The Scenery of Scotland* was also an important benchmark in the field. As if to demonstrate his versatility, he also published in two volumes in 1897 *The Ancient Volcanoes of Great Britain*, which was the first comprehensive study of its kind of any country in the world. Geikie's work was later built upon by others in the Geological Survey including Sir Edward Bailey and C.T. Clough. The study of igneous rocks remains an active area for research by academics in Scotland and internationally.

'ROCKS FROM VOLCANIC FIRES' – SINCE GEOLOGICAL RECORDS BEGAN

Although there are no active volcanoes in Scotland today, the Scottish geological record is littered with igneous rocks comprising a wide variety of types. The dates when each volcano was active can also be established by careful study. In the previous chapter we learned about igneous rocks interspersed with Lewisian gneisses found on the Isle of Harris and in the Northwest Highlands. Since the geological record began in Scotland, igneous rocks have been an ever-present feature. It is an impossible task to describe each and every volcanic episode that followed, even in the scantest of detail. So this account will be nothing more than some highlights, with a focus (in the next six sections) on Scotland's most recent and pervasive volcanic episodes, which took place (some 60 million years ago!) during Eocene times. But before we get there, let's canter through the best of the rest.

As continents collided – we have already seen that the Iapetus Ocean closed about 420 million years ago, and the continents of Laurentia and Avalonia collided, creating the mountains of Scotland. This event is called the Caledonian Orogeny (orogeny meaning 'mountain-building event', normally caused by colliding continents) and it gave rise to igneous rocks of many different types. Some were formed at depth whilst others were erupted as lavas at the surface.

Let's consider some of that variety:

- Granites formed at depths as the continents ploughed into each other, causing the crust to thicken. At the base of the newly thickened crust, temperatures and pressures were unimaginably high, so the lower crust melted and copious amounts of granite were created in this crucible-like environ-

Sir Archibald Geikie

Cairngorm granite has been deeply scarred by the erosive force of the ice. These great scoops out of the bedrock are known as the Northern Corries and are well-known features that bound the Cairngorm plateaux.

ment. We know that these rocks, now at the surface, must have formed at considerable depths in the Earth's crust. The crystals that comprise these granites are large. Slow cooling at depth in the Earth's crust is required to form crystals that can be seen without a hand lens. Again erosion is the reason that caused these rocks to be exposed at the surface. Granites generated at the time of the closure of the Iapetus Ocean are very widespread throughout Scotland. Most of the red blobs on the geological map of Scotland, with the exception of those on the Hebridean islands, are of this age. It's an extensive catalogue of iconic places in Scotland that are built from this rock. They includes Cnoc na Maole (Helmsdale granite), Lochnagar, Ben Macdui, Hill of Fare, Strontian, the hills around Loch Etive, Loch Doon, Cairnsmore of Fleet and Cheviot, which is shared with Northumbria.

- Gabbro is similar to granite in being coarse-grained, but it is chemically very different and much darker in colour. In molten form, it is the magma from which basalt originates. The distribution of intrusions of gabbro formed during the Caledonian mountain building is largely restricted to the northeast of Scotland. Gabbros don't give rise to extravagant landscape features, like the Cairngorms for example. Morven, north of Ballater, and the area around Inch are perhaps some of the best places to inspect these rocks. They are most readily identified by the distribution of aggregate quarries, as they are an ideal material for use as road stone.

Ben Nevis towers over the town of Fort William and Loch Linnhe

- Ben Nevis (Britain's highest mountain) and Glencoe were ancient volcanoes that once rocked the world. We haven't heard a peep from them for over 400 million years, but when they roared into life just after the Iapetus Ocean closed, they were super-volcanoes. They are also of great historical interest in the development of our understanding of how volcanoes work. Glencoe was a monster by any standards – a 14km by 8km cauldron of boiling, bubbling and erupting rock. After one of the early eruptions, the volcano collapsed in on itself, creating what we know as a caldera, releasing pressure and causing vast amounts of superheated gas and ash to erupt. This style of volcanic eruption was first recognised by the Geological Survey's Sir Edward Bailey. He shared his findings with colleagues in the Geological Society of London shortly after he had his 'eureka moment'! It is said he travelled directly from Glencoe to London, still wearing his heavy boots and other field gear, so keen was he to tell of his new ideas. Pyroclastic flows, similar to those that killed the inhabitants of Pompeii two millennia ago, have also been identified at Glencoe, so we can imagine the ferocity of the eruptions that swept across these early landscapes.
- The cold dead remnants of countless ancient volcanoes litter the Scottish Borders. These volcanic corpses take the form of either circular plugs of cooled magma that mark the site of the ancient throat of the structure, or the associated lavas that flowed from the volcanoes during one of their many

Hareheugh Crags is one of many ancient volcanic stumps that punctuate the Scottish Borders. This is the core of the ancient volcano with the associated lava field spread over many square kilometres around this central feature.

eruptions. Hareheugh Crags near Kelso has both parts of this volcano story. A prominent crag that can be seen for miles around marks the central pipe along which the magma flowed upwards from a deep molten chamber. The rich agricultural land to the west, east and south is underlain by the lavas that belched from the Hareheugh vent. Basalt is by far the commonest type of lava found throughout the world. It can be divided into a series of sub-types by detailed study of the constituent mineral content. One such variant of basalt is known as *hawaiite* – first identified on the island of that name. The rocks at Hareheugh share the same mineral characteristics as the lavas currently being erupted from the active volcanoes on Hawaii and have been recognised as such for the best part of a century.

Volcanoes of the Inner Hebridean Islands

Before the flares of the Inner Hebridean volcanoes lit up the night sky, Scotland was part of a major landmass called Pangaea – from the Ancient Greek *pan* and *Gaia*, meaning *All Earth*.

But soon cracks started to appear. The separation of Europe from North America took place as rifts appeared either side of the British Isles.

In the area now recognised as Iceland, this process of disintegration was accelerated by the development of a ferocious font of energy, known as a mantle plume. Rising from deep in the planet, it acted like a blow-torch, focusing its fire on the base of the Earth's crust.

Most of the islands of the Inner Hebrides of Scotland were formed as part of the volcanic process that created the North

VOLCANIC FIRES

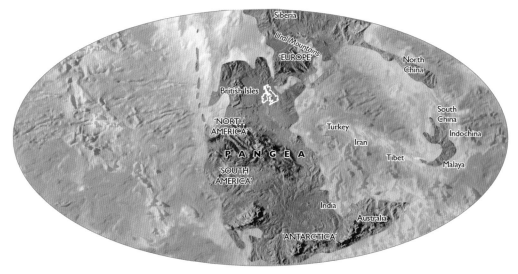

Above.
This is a reconstruction of the ancient continents as they were 250 million years ago during Triassic times. The continents had largely coalesced into a single landmass spanning the globe from pole to pole, called Pangaea. The land that was to become Scotland had some strange bedfellows at this time – North America perhaps being the most surprising. But any configuration of land and sea didn't stay fixed for long, and the continents were soon on the move, powered by the churning motions of the underlying mantle.

Left.
Pangaea breaks up. A proto-Atlantic Ocean formed. Prompted by the writhing currents in the upper mantle, great cracks appeared in this super-continent, and most of the continents that we now recognise came into being.

Atlantic Ocean. The islands of Skye, Rum, Eigg, Canna, Mull, Arran, Ailsa Craig and the western outpost of St Kilda all owe their origin to this violent volcanic past. Again, the classic volcano shape is long gone, as in every case the volcanic cone and associated lava fields have been ground down and partly eroded by the passage of ice and time. But the unmistakable elements that are characteristic of a volcano remain. In fact, erosion has enhanced our view of these magnificent edifices. Ice, wind and water have cut deep to the very heart of these structures, so that their inner plumbing is now on show at the surface.

Skye – the volcanic isle

In another first for British scientists, the pioneering Victorian geologist Dr Alfred Harker published a seminal account of the geology and associated mineralogy of the ancient Skye volcano. Harker was a graduate of St John's College Cambridge and was employed by the Geological Survey of Great Britain to undertake a geological survey of the island of Skye. Harker's new insights on the inner workings of the Skye volcano were made available as a 'one inch to the mile' geological map, and the accompanying classic Survey Memoir entitled 'The Tertiary Igneous Rocks of Skye' was first published in 1904.

The majestic peaks of the Isle of Skye are unmistakably the eroded stumps of a massive volcano that first became active around 60 million years ago. The Black Cuillin are chilled remnants of the magma chamber that powered this volcanic system. Harker's account provided the first detailed and systematic account of the geology and associated mineralogy of a long-silent volcano and prompted similar studies to be undertaken in many other parts of the world.

The Black and Red Cuillin are made from very different stuff. They are the solidified remains of the magma chambers that fed the volcanoes on the surface. The black coloration of the coarse-gained rocks that made up the majestic peaks of Sgurr Alasdair, Sgurr Dearg, Sgurr nan Eag, and many more besides, comes from minerals that accumulated layer upon layer on the floor of the magma chamber of the Skye volcano. Many layers of the minerals pyroxene and olivine built up like layers of sand and mud on the ocean floor to form a wide variety of basic and ultrabasic rocks with exotic-sounding names such as gabbro, eucrite, allivalite and peridotite. They were frozen in time as the volcano went silent and were later revealed at the surface as erosion stripped away the upper layers of the structure.

In contrast, the Red Cuillin comprises granite formed when the intense heat of the volcano melted the basement rocks. Great balloons of molten rock, less dense than that of the surrounding strata, rose through the Earth's crust to a position close to the surface. With the passage of time, the layers of rock that separated these balloons or intrusions of granites from the surface were stripped away by erosion. The resultant 'unroofed' granites now add to the eclectic mix of rock seen at the surface.

The Black Cuillin (top) and the Red Cuillin (left) of Skye comprise some of the finest scenery Scotland has to offer.

Skye is classic ground for the study of volcanoes. The dramatic sea cliffs and rugged peaks provide abundant exposure of the rocks of the ancient volcano. The Trotternish Peninsula, running due north from the central mountainous core of the island, is built largely from lavas that were erupted from a fissure or series of deep-seated cracks in the Earth's surface at an early stage in the volcano's development. Individual lava flows can be traced along the Trotternish Ridge that are consistent in thickness and mineral content. As today, eruption of an individual lava flow is a rapid event. Each lava was succeeded by the next flow after an elapse of time. So the upper surface of each flow was subjected to the elements before the next flow swept across the land. Study of the tops of these lava flows and indeed the sediments that accumulated between each eruption has told us a great deal about the climate of this part of Scotland as it existed 60 million years ago. It may be difficult to comprehend on a wet and dreich Wednesday in Portree that these lavas were erupted into a tropical

The Trotternish Penninsula on Skye is built from layer upon layer of basalt lava that seeped from the Skye volcano. Two large landslips formed which destabalised the lava pile – the Quiraing (illustrated in this photograph) and the Old Man of Storr, (see frontispiece). Both features are relatively recent landslips that occurred after the ice melted over 10,000 years ago.

world where ginkgos and other exotic species of tree clothed the land during the short periods between the eruptions. We know this because of fossilised plant fragments found in the sediments that were preserved between the lavas and also because the tops of the flows have been deeply weathered in a tropical climate.

The lava pile rests on a thick platform of sediments of Jurassic age. Many layers of limestones, shales and sandstones were deposited around 160 million years ago in shallow seas and coastal lagoons that existed here at that time.

Rum and the Small Isles

The adjacent islands of Rum and the Small Isles also have a volcanic origin. Rum was a separate and equally energetic volcano that erupted over a period of two million years. But there is a twist to the tale of the Rum volcano. During the relatively short period it was active, it collapsed in on itself early in its development to create a caldera. A central core of rock subsided into the molten magma chamber below as the volcano blew its top during a particularly violent eruption. Perhaps the best known contemporary caldera volcano is Krakatoa in Indonesia. This volcano erupted last in 1883, killing many and sending shockwaves around the world – literally. Global temperatures dropped as a result of the ash cloud generated by the eruption. It is possible that the eruption of the Rum volcano could have made its mark in a similar fashion. What is particularly remarkable about the

Rum volcano is the preservation of the magma chamber, laid bare by erosion. The highest peaks of the island, Askival and Hallival, are made from layer upon layer of igneous rock deposited on the floor of the magma chamber. This volcano played an important role in the understanding of how the processes of crystal settling worked in ancient and, by implication, contemporary volcanoes. Boat-loads of geologists have visited Rum over the years to examine these classic geological features.

The Sgurr of Eigg, on the neighbouring island of Eigg, is a dramatic ridge of columnar rock that dominates the skyline. It provides an intriguing glimpse of the final phases of volcanic activity in this part of Scotland. Many layers of basalt lava had already been spewed out across the ancient landscape, creating a barren and inhospitable place. Next came a calmer interlude when the eruptions stopped and a wide river flowed, dumping rounded pebbles and layers of sand that were later squashed to form a deposit known as a conglomerate or pudding stone. A black, sticky lava of unusual composition, known as a pitchstone, was later erupted from a nearby volcano, and this river of viscous molten rock inched its way across the landscape, burying the layers of conglomerate. Over time, the lava cooled and formed impressive pipe-like columns many tens of metres high that are a key element of the signature views of this beautiful island.

The island of Canna, lying to the northwest of Rum also has an interesting story to tell. It is part of an extensive layer-cake of lavas that run north and eastwards towards Skye and so in all probability were erupted from the Skye volcano. What makes Canna so special are the many layers of water-lain boulders and associated layers of sand that were dumped by braided streams and rivers that flowed across the bare rock between eruptions. Some of the boulders are over a metre in diameter, indicating the strength of the river currents that transported them to their

A spectacular sunset over the Isle of Rum highlights the main peaks of Askival and Hallival. They are eroded remnants of the once mighty Rum volcano.

The Sgurr of Eigg dominates the skyline. Events that shaped this island represent one of the final phases of volcanic activity in this part of Scotland. The lava that builds the upper part of the Sgurr cooled to create pipe-like structures, similar in form to the rocks at Fingal's Cave on Staffa. A fuller description is given overleaf.

present position. Pebbles from these deposits have been matched with the bedrock of parts of Skye, so there is good evidence that the rivers flowed from the north.

Ardnamurchan Point

Continuing our march southwards, Ardnamurchan Point is our next destination. This promontory is a long, bony finger of land that pokes into the Atlantic Ocean, forming the most westerly point of the British mainland. The western tip is one of the most singular landscapes in Britain. The volcano that built this place has been stripped down to the basement level revealing an amazing arrangement of rock. Concentric circles of magma are frozen in time, pulsed from a deep source. Each of these ring dykes was formed as pressure built up from below, creating a crack in the Earth's crust that was circular in plan. Molten rock then seeped to the surface as the central core of the volcano foundered into the magma chamber below. This process was repeated a number of times to create this striking landscape. We also see how the focus for volcanic activity can shift over time. Three separate volcanic centres can be distinguished here, with the second cross-cutting the structures created in the initial burst of volcanic activity and the third cutting across traces of both earlier structures. The Ardnamurchan volcano then fell silent, so the grain of the landscape largely follows the third and climactic

To describe the rocks of Adnamurchan Point as one of the wonders of Scotland isn't too much of a geologist's flight of fancy. It is amazing to see the magma chamber of a once active volcano exposed at the surface. It also confirms the unimaginably erosive power of the ice that scoured the land and, over millennia, sliced away the upper part of this once-towering volcanic edifice.

phase of events. From the air, these ring dykes are one of the wonders of Scotland.

Isle of Mull

The island of Mull lies to the south of Ardnamurchan Point. It too owes its origins to the opening of the Atlantic Ocean. It was yet another volcanic boil that blistered and burst, part of the chain of active vents that defined the western edge of Scotland. As with the Rum volcano to the north, it was also a caldera that blew its top as pressure built up from below. Magma accumulated beneath a thick cap of lavas that had already been erupted. Pressure mounted as the rising streams of molten rock could not make it to the

This view of the lavas of the Ardmeanach Peninsula, north of Loch Scridain on the Isle of Mull, is an impressive sight. Each step in the staircase that builds this hillside represents a separate volcanic event where great gushes of runny magma were erupted onto the Earth's surface and flowed across the barren landscape until they solidified. Successive lava flows were stacked high as the volcano erupted time after time to create a pile many kilometres thick.

surface and had nowhere else to go. The 'pressure cooker' eventually blew, creating a deep depression at the surface and shockwaves that would have been felt on the other side of the world.

Arran and Ailsa Craig

Our final destination in this whistle-stop tour of the volcanic islands of Scotland's west coast is the island of Arran and the adjacent volcanic sentinel of Ailsa Craig. Arran is often called 'Scotland in miniature' because it has the emblematic elements of highland and lowland. The geology, too, is a microcosm of the country as a whole with ancient basement rocks overlain by younger sediments. The Highland Boundary Fault sliced through the island over 400 million years ago, and now separates the higher ground from the more undulating productive farmland to the south. And we have evidence for another volcano. Goat Fell, which isn't quite a Munro at 2,866 feet, dominates the granite hills of northern Arran. In plan, the highest ground on the island is underlain by an almost perfect circle of granite, part of an igneous intrusion that punched its way to its present position through a cover of much older rocks. As with every other volcano

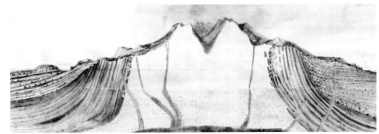

Top.
This east-west geological section through of the island of Arran was prepared by John Clerk of Eldin in 1787. It is hugely insightful and shows remarkable understanding at a time when geology was still in its infancy. The central core of Arran is built from once-molten granite that pushed up from below, displacing the older layers of sedimentary rock in the process.

Bottom.
Most of the volcanoes described in the preceding pages had an associated swarm of linear pulses of magma or 'dykes' that radiated from the central volcanic core. These features are usually just a few metres across, but run largely in a straight line, in some instances, for many hundreds of kilometres. The south coast of Arran is a particularly good place to see these features, and this photo shows the nature of one particular dyke as it heads out to sea.

of this vintage in Scotland, the characteristic cone has been ground down by wind, water and ice to foundation level and the ancient magma chamber has been laid bare. This was fertile ground for the early geological pioneer James Hutton, who visited Arran in 1787. He and his travelling companion John Clerk of Eldin deduced the underlying structure of the Northern Mountains of Arran from the distribution of rocks at the surface and produced one of the first geological cross-sections ever published. It showed the granite thrust upwards from the lower reaches of the crust into its present position with younger strata displaced and tilted upwards as the granite inched into place. This interpretation has not been bettered to this day.

A common feature of all of the ancient volcanoes of the Scottish islands is the occurrence of volcanic stripes across the countryside, known as dykes. These linear features were created when molten rock was squirted from the main volcanic centres, often in a radial pattern, into the surrounding rock. The pressures involved must have been enormous as some dykes have been found many hundreds of kilometres away from the volcano that produced them. Most dykes are between 2 and 3 metres wide, but can be considerably wider.

And so to Ailsa Craig. This granite plug is to the sport of curling what a stand of willow trees is to cricket. Thousands of curling stones have been hewn from this thumb of granite that rises abruptly from the steely-grey waters of the Firth of Clyde.

Our journey through the volcanic islands of Scotland is complete. We have traversed the leading edge of the crust that was pulled apart as Scotland was split asunder from North America. These islands are enduring evidence of the events that helped shape the landscape of Scotland.

Ailsa Craig sits due west from Girvan – a solitary, disconnected granite pile. As with many of the other volcanoes of this vintage, much of the superstructure has been removed by the passage of ice, and in the case of Ailsa Craig, all that remains is the solidified magma chamber.

CHAPTER 4

Life on Earth

Fossil remains of past life are our only window on the diversity of plants and animals that existed during Planet Earth's long history. Worldwide, around 99 per cent of all creatures that have ever lived on this planet are now extinct, so the study of fossils, where we are able to recover them from the rocks, is very instructive. Geologists have discovered and described whole dynasties of exotic long-extinct creatures, such as the dinosaurs, on the basis of their fossilised bones and the rocks from which their remains were recovered. The process of fossilisation is serendipitous, almost random, as it is reliant on ideal conditions for preservation. So our picture of past life is incomplete and represents only a few meagre scenes from the epic pageant of life on Earth since earliest times. Animals with hard parts, such as shells and bones, are disproportionately represented in the fossil record, whereas organisms such as jellyfish and delicate plants usually vanish without leaving a trace. But the incomplete, fragmentary

Hydrothermal vents or 'black smokers' are created where tectonic plates move apart to form new crust. They are typically located at the midpoint of all of the world's oceans, along a topographic rise, known as a mid-ocean ridge. Fuelled by the associated molten rock, water emerges from these vents at temperatures in excess of 450°C. The waters are normally loaded with sulphides and other toxic chemicals, which gives rise to the black coloration. Some scientists believe that life on Earth originated in this extreme environment.

When the planet was newly formed, the Earth's crust was unstable and pockmarked with active volcanoes. The early atmosphere was largely derived from gases belched from these erupting volcanoes. In the early years of the planet's existence, life-sustaining oxygen was absent from the atmosphere. It took billions of years from these earliest times before the planet was able to support even the simplest life-forms.

and unrepresentative fossil record is all we have to work with. In line with all our other scientific endeavours in the field of geology, we have made the most of what is available.

For billions of years after the Big Bang and subsequently when our planet was a ball of cooling rock, the Earth was entirely devoid of life. Nothing, not even the tiniest single-celled organism, could withstand the harsh conditions on the surface of the Earth. Some scientists think that first life may have come from outer space aboard a comet. Others speculate that the volcanic vents comprising the spine of the mid-ocean ridges (also known as black smokers or hydrothermal vents) are the most likely place where the first viable seeds of life – the cell – were energised.

We won't delve too far into the mystery of how life came to be in the first instance, but, at some point in our distant geological past, the transition from the chemical building-blocks for life to the first viable life-forms was clearly made.

In those earliest times, the Earth's atmosphere contained no oxygen, only nitrogen, carbon dioxide and methane. The atmosphere was derived from gases erupted from the many volcanoes that were active at the time.

The high levels of greenhouse gases in the atmosphere ensured that the Earth's temperature was much higher than today – somewhere between 55°C and 80°C. Only very specialist, primitive organisms could withstand such conditions. Over time, the atmosphere became more benign, and life on Earth began to flourish. It was from small beginnings that more complex and diverse ecosystems started to develop.

The earliest trace of life in rocks from Scotland dates back one billion years. The red sandstones of the Torridonian hills of the Northwest Highlands yielded an enigma – a microfossil thought to be an early type of algae. Named *Torridonophycus*, it is probably a green alga.

Some of these early life-forms had one remarkable quality – they were net producers of oxygen as a by-product of their lifestyle, so these primitive organisms changed the atmosphere of the early planet. Subsequently, individual cells of organisms that evolved from these early life-forms became bigger and more specialised. This set the scene for what has become known as the 'explosion of life' during the later Cambrian Period. This phenomenon is the most significant event in the history of life on Earth. From a few primitive algae during earlier times, the oceans began to teem with life of many different types. At this time, life was entirely confined to the oceans. The invasion of the land by plants and animals came later. Creatures with shells, and perhaps most significantly of all with backbones, appeared for the first time in the worldwide fossil record. The stretch of geological time from this point on to the present day is known as the Phanero-

Burgess Shales

The Burgess Shales in Canada illustrate the 'explosion of life' better than anywhere else in the world. From the single-celled organisms of earlier times sprang an extraordinary diversity of life as captured in these shales. What is most unusual is the wide variety of soft-bodied animals that have been preserved for posterity. Some of nature's evolutionary experiments – such as the beautifully named *Hallucigenia* with its row of spines and large nose-like structure at one end, or the amazing *Opabinia* which had five prominent eyes and a long trunk-like appendage that had small spines at the end for catching prey – were bizarre design concepts that were never going to be successful and, of course, they were not. These early prototypes co-existed alongside hard-shelled creatures, such as bivalves and trilobites, that were common in many of the world's oceans.

Fossils from the Burgess Shales: *Marrella splendens*, an anthropod (left) and the shrimp-like *Anomalocaris* (right)

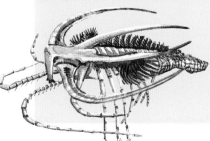
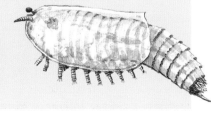

zoic. This is the last 545 million years of Earth history, when life was abundant on Planet Earth. The remains of these early, and in many instances fast-changing, lineages of plants and animals are captured in the fossil record.

FIVE MASS EXTINCTIONS

Although our focus is largely on what happened in Scotland, we can't isolate ourselves from past events of worldwide significance. In terms of life on Earth, perhaps the most significant of these events were the five occasions when life was almost extinguished or suffered catastrophic decline. There are some who think we stand on the cusp of a sixth 'age of dying', where the species *Homo sapiens* (i.e. us) has played a central role in the demise of many plant and animal groups.

The first mass extinction happened at the end of the Ordovician Period, some 444 million years ago. The second occurred at the end of the Devonian Period around 359 million years ago; the third at the end of the Permian around 251 million years ago; and the fourth at the end of the Triassic some 200 million years ago. But perhaps the most celebrated is the mass extinction event that took place at the end of the Cretaceous Period, about 65 million years ago. It was at this point that the dinosaurs and flying pterosaurs became extinct.

All of these events significantly impoverished life on Earth, both in terms of species that were lost at a geological stroke and also in terms of the number of ecological niches that were left temporarily vacant. The cause of these events has inevitably been a rich source of speculation: meteor strikes from outer space, volcanic eruptions that affect the composition of the atmosphere, acid rain, ozone depletion, changes in ocean circulation patterns and long-term global warming are all potential culprits for these catastrophic events, either singly or in combination. But life on Earth is resilient. Ecosystems adapted and re-established themselves after these catastrophic events, and life continued.

SCOTLAND'S DISAPPEARED ECOSYSTEMS: CHANGING ENVIRONMENTS

Ecosystem is the term used to describe a biological community of plants and animals interacting with the physical environment. The

Caledonian pine forests and a sand dune–foreshore area are good examples of contemporary ecosystems. Let's consider life on Scotland's tiny fragment of Planet Earth as a series of ancient, long-disappeared and ever-changing ecosystems. Each evolved in response to where the 'land that was to become Scotland' was on its global journey at any particular time and the environmental conditions that developed as a consequence. A wide variety of environments are recognised that record this epic odyssey across the globe.

From before Cambrian times and for a period of around 200 million years, the deep dark waters of the Iapetus Ocean were home to the first animals recorded in our fossil record. Primitive plants 'came ashore' during the late Ordovician and diversified considerably thereafter. During Devonian time, life continued to evolve to the extent that both plants and animals could thrive on land and in the associated freshwater lakes. The lake environments were replaced by verdant tropical rainforests of the Carboniferous Period when Scotland sat astride the Equator. An associated shallow-water reef ecosystem also forms part of the fossil record from this time. Drifting further north to the latitude of present-day Saharan Africa, desert conditions prevailed and a thick carpet of red sandstones date from the Triassic and Permian Periods. During Cretaceous times when sub-tropical seas covered most of the country and thick layers of chalk were deposited in southern England, Scotland was also largely submerged under these shallow tropical waters. Strangely little trace is to be found of this episode today. The succeeding Palaeogene Period left its mark in terms of the explosive volcanic activity described in the previous chapter. But despite the violence of these volcanic events, the remnants of sub-tropical ecosystems that blossomed between the episodes of volcanic activity are preserved in the layers of sediment that lie between successive lava flows. The final ecosystem is that of the tundra with Arctic-like conditions prevailing during the Ice Age. It is at this point that we note the first evidence for the presence of our ancestors.

But, as the final chapter explains, the ice will return in the future and tundra conditions or a covering of ice and snow will once again be re-established across what is presently a green and pleasant land. But more of that story later.

Let's now focus in more detail on the main ecosystems that have supported life in Scotland since geological records began.

LIFE ON EARTH

Life under the ocean waves

During Cambrian times, the land that was to become Scotland was a little piece of the North American continent. England, Wales and the rest of Europe was on the other side of the Iapetus Ocean.

The Iapetus Ocean played a huge role in the creation of the bedrock of Scotland. Many of the rocks that were later to form the Highlands and Southern Uplands were laid down as sands, silts, clays, muds and limestone reefs in this expanse of water that became, at one point, wider than the present-day North Atlantic Ocean. For a period of around 200 million years, this ocean played host to the animals that were eventually to turn up as fossilised remains in the rocks that now lie within Scotland's borders. During this time, no fossils from land-living animals are recorded. Our legacy of fossils from the Cambrian, Ordovician and Silurian Period are therefore entirely related to the deep ocean or the habitats that existed around the fringes of the Iapteus Ocean.

Fossils of the Cambrian Period

Some creatures 'disappear' completely when they die, but traces of their former existence may still remain in the fossil record in the form of their footprints, burrows, bite marks left by predators or their poo! In this case, it's worm burrows in beach sands that have been preserved. The top picture is a section through the sands and the lower picture is a plan view of the circular nature of each of the burrows. But the creature that made these structures had no hard parts in its anatomy, so its remains were not preserved.

Trilobites are small woodlouse-looking creatures that inhabited the ocean floor. These examples of the *Olenellus* trilobite were first described over 100 years ago from shales in the North-west Highlands.

Fossils of the Ordovician Period

During this time, there was a phenomenal proliferation of all forms of marine life worldwide. Trilobites continued to survive and prosper into the Ordovician and were joined by bryozoans (soft-bodied animals that lived within a hard skeleton of calcite), brachiopods (bottom-dwelling shelled creatures), bivalves (filter-feeding creatures with articulated or joined shells) and a hugely significant animal group – the graptolites. These enigmatic colonial animals floated in the ocean, tossed this way and that by the vagaries of the ocean currents. Graptolites evolved into many different forms including single branches, double branches and spiral habits. They proved to be hugely valuable in providing relative dates of rock sequences from different areas of Scotland, indeed throughout the world. The end-Ordovician mass extinction saw biodiversity in the world's oceans greatly diminished.

Fossil of Ordovician age – *Leptaena* – from near Girvan

Fossils of the Silurian Period

By late Silurian times, the Iapetus Ocean was no more. The various elements of land had come together to form what would later be recognised as Britain. It formed part of the larger continent of Laurentia that lay south of the Equator. In terms of life on Earth, one of the most significant developments that took place during Silurian times was that plants started to colonise the land. These early invaders consisted of simple leafless structures, not much bigger than a pin. In the UK, the evidence for this colonisation comes from fossils found in Wales. The specimens recovered are not showy or impressive to look at, but the significance of this transition from ocean to the land is massive. This is also the first time that fish appeared in the fossil record in Scotland. These early creatures had no jaws, but possessed a sucker mouth and a row of gills for breathing. Rejoicing in the name *Jamoytius kerwoodi* these early fish are related to groups that swim in our waters today – the hagfish and lampreys. Early arthropods, called *Eurypterids*, and more commonly known as 'sea scorpions', have also been found from rocks of this age. These creatures were carnivores and probably preyed on other arthropods and fish.

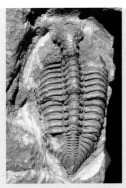

Above: a fine specimen of the trilobite *Encrinurus* recovered from the rocks of the Pentland Hills; and (right) worm burrows (but not including the animals themselves) from muds deposited in the deep waters of the Iapetus Ocean. This rock slab was recovered from near Innerleithen.

LIFE ON EARTH   55

The age of the fishes and the rise of plants with hot springs thrown in!

With the Iapetus Ocean now closed, the range of habitats available to life on the land that was to become Scotland radically diversified. This fragment of the crust was now largely landlocked, sitting between the towering mountains created by the closure of the Iapetus Ocean. A great freshwater lake, Lake Orcadie, with a sporadic connection to the sea, formed in this arid and baked land. It teemed with primitive heavily armoured fish, and plants continued to get a firmer hold on the adjacent dry land. Living rock-like structures, known as stromatolites, also thrived in the clear shallows around the edge of the lake.

We know all of this because of the thick band of layers of sands and muds (around 4 kilometres in total) that built up on the lake bed and, of course, because of the fossils entombed by these sediments.

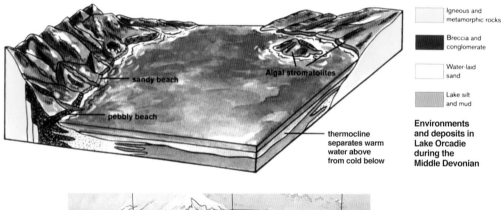

Reconstruction of what the environment might have looked like some 400 million years ago. This great freshwater lake covered much of what we know as the Moray Firth, Black Isle, Caithness, Orkney and the Shetland Isles. Red-coloured sediments from this time are still widely seen in these areas.

Environments and deposits in Lake Orcadie during the Middle Devonian

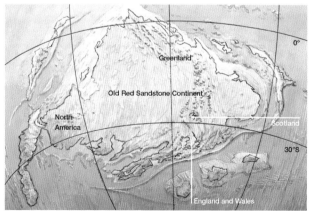

The Old Red Sandstone continent – this new landmass was created by the closure of the Iapetus Ocean. It included what we now recognise as North America, Great Britain and Ireland and much of Europe

Fossils from the Devonian (also known as Old Red Sandstone) Period

Left.
The Devonian Period is known as the 'Age of Fishes'. Many different genera (types) of fish have been recovered from the sandstones of this age. The middle image is of a specimen of fossil fish called *Pterichthyodes milleri*. An image of this fish has been cut into a paving slab at Hugh Miller's cottage in Cromarty (top). The bottom image is the best guess at what this creature would have looked like in life.

Right.
Hugh Miller

Fish – the freshwater lakes of this time were home to a wide variety of fish. Most are preserved in discrete beds of rock, known as 'fish beds'. It was during the times these layers were laid down in the lake that conditions for fossilisation were ideal. The fossilised fish remains were first described in print by Professor Roderick Impey Murchison, then Director of the Geological Survey, who visited Caithness in the 1820s. Since then, many have studied the area and frequent discoveries new to science have been made. Perhaps the most famous discoveries were those made by the Victorian scientist Hugh Miller. Born in the village of Cromarty on the Black Isle in 1802, he worked as a stone mason for over fifteen years. His work in local quarries exposed him to the variety of past life recovered from the rocks. Miller was one of the first scientists who tried to popularise geology and take it beyond the narrow confines of the scientific community. He said 'Let me qualify myself to stand as interpreter between nature and the public.' His efforts are immortalised in the naming of a fossil fish, *Pterichthyodes milleri*, which is commonly found in the sedimentary rocks of Miller's home area. The fish has body armour in the form of arrow-shaped scales and short fins either side of its body that look like wings. Other fish from this era include *Dipterus*, also known as the lung fish, whose distant relatives still exist today. It too was covered in tough scales and had a set of robust teeth that were probably used for crushing shellfish.

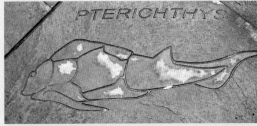

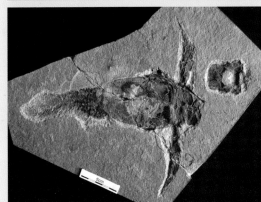

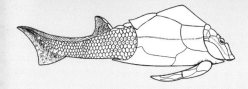

Plants and spiders – this was the time in geological history when the land turned green. The tentative toe-hold that plants had on land in earlier times was further exploited with an explosion of plant diversity. Club mosses, horsetails and primitive seed plants made their appearance during Devonian times, all probably derived from a common ancestor. Many new ecosystems developed on land as a result. Pre-eminent amongst the fossil plant heritage in Scotland are the exquisitely preserved specimens from the rocks near the village of Rhynie in Aberdeenshire. Study and interpretation of the rocks buried beneath a tranquil field near Rhynie have propelled this place to a status of a site of international palaeobotanical importance. The Rhynie Chert is an ancient ecosystem that developed around a hot spring, comprising some of the world's earliest plants, fungi and many long-extinct spiders and other arthropods. The preservation of elements of the spider remains is so good in the Rhynie Chert that even some soft parts of these tiny creatures, such as the lungs, have been preserved. The quartz-rich spray from the hot spring coated and eventually buried the adjacent plants and animals, and the ecosystem was then frozen in stone. That was the last time it saw the light of day until excavated some 400 million years later.

Above.
A reconstruction of the *Palaeocharinus* spider was found entombed in the Rhynie Chert.

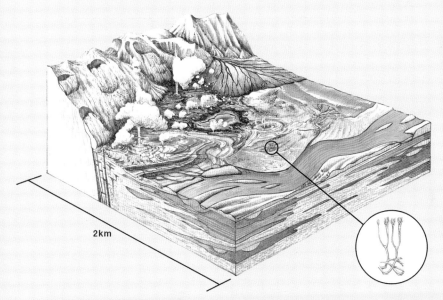

Left.
A reconstruction of the Rhynie ecosystem

Stromatolites – these simple organisms, a type of cyanobacteria, lived in colonies in clear shallow water around the lake and grew to form rock-like structures. They persist to this day, but are not to be found in Scottish waters. They are some of the oldest evidence of life to be found on Earth. In Scotland, the earliest examples date from the Precambrian rocks of Islay, but excellent specimens have also been found in rocks of Devonian age from Orkney. This specimen, from the National Museums of Scotland collection, shows layers that built up year on year, as the colony grew. Stromatolites were one of the group of organisms that produced oxygen as part of their natural cycle by a process known as photosynthesis. In so doing, they supported the development of other oxygen-breathing life-forms.

Right.
Stromatolites

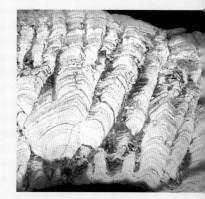

Worldwide, the end of the Devonian Period saw another mass extinction. Again the cause is a matter for debate, but one interesting idea is that the circulation of ocean currents changed dramatically, and stagnant bottom waters may have been disturbed, poisoning the upper levels of the water column and also coastal areas where most of the invertebrates and other ocean-dwellers lived.

Tropical rainforests and shelf seas

By the time of the Carboniferous Period, plants and animals had diversified into myriad forms. Life – both plants and animals – was now firmly established on land and in the water. 'Scotland' sat close to the Equator at this time. Tropical rainforests and warm shallow seas that lapped against the land were the predominant ecosystems.

From this time in the Carboniferous Period, plants began to grow to a towering size: 40m in height according to some estimates. Forests of *Lepidodendron* covered the land and supported a diverse ecosystem of other plants and animals. Eleven fossilised *Lepidodendron* tree stumps of Lower Carboniferous age are beautifully preserved in Victoria Park in Glasgow. An ornate pavilion was erected in Victorian times to preserve these amazing fossils for posterity. This perhaps marks the earliest act to conserve our fossil heritage. This site is all the more remarkable as it is a virtually intact fossilised fragment of forest that is about 325 million years old.

One of Scotland's most famous fossils is 'Lizzie the Lizard' or, to give it its scientific name, *Westlothiana lizziae*. Its remains are a prominent exhibit in the National Museums of Scotland in

Coal

The tropical rainforests of the Carboniferous gave rise to coal. This mineral powered the Industrial Revolution in Scotland and sustained the tight-knit mining communities that sprang up wherever this resource was exploited. This ancient ecosystem, or its fragmentary remains, was therefore a key determinant of the way we lived our lives over 300 million years later! Coal seams and their attendant deposits are the most intensively studied geological deposit in the country as their exploitation has given rise to huge economic benefits, albeit with an environmental downside that we are just beginning to come to terms with.

Reconstruction of a coal forest ecosystem with a dragonfly (a), *Lepidodendron* (b), tree fern (c), *Calamites* or horse-tail (d) and amphibian (e).

Fossilised *Lepidodendron* tree stumps of Lower Carboniferous age in Victoria Park, Glasgow

Edinburgh where the story of its discovery, excavation and the appeal to the nation for its purchase is told in one of the display cabinets. Its nickname is slightly inaccurate as this creature is no longer regarded as having all the characteristics of a lizard. But it was certainly a four-legged lizard-like creature that inhabited the steaming tropical forest and lake environment that was to become West Lothian 320 million years later (see illustration overleaf).

The remains of insects are rare in the fossil record. So when good specimens are found, they are pored over by experts across the decades. The remains of the rare dragonfly *Idoptilus* were first described in 1887 and are still the subject of study today (see illustration overleaf). The latest view is that this find represents the nymph stage of a dragonfly that existed in Carboniferous times. Oxygen levels were higher in the atmosphere at this time than the present day, so insects, in particular, grew to extravagant sizes. Some dragonflies had a wingspan of half a metre or more. Other insects recovered from the fossil record of this time include a shrimp-like creature on which the delicate antennae and legs have

This 'shackle of bones' was an amazing find that intrigued the scientific world. This specimen, soon to become known as 'Lizzie the Lizard', was found by Stan Wood, a professional fossil hunter.

been beautifully preserved, the remains of fossil scorpions up to 90cm in length, as well as tiny mites and harvestmen.

Lapping against the shores of the tropical rainforest was another world that teemed with life – a tropical sea. This marine world supported sharks up to 5m in length. *Akmonistion zangerli* was an odd-looking beast that carried a strange dish-like structure on its back. It was unearthed in Bearsden in Glasgow, one of the few sites in the world where sharks of this age have been found. The warm shallow seas of this ancient tropical environment also supported an abundance of corals, gastropods (molluscs with coiled shells) and fish.

Fossilised remains of the dragonfly *Idoptilus*. The story of this fossil is told overleaf.

The National Museums of Scotland created this reconstruction of what a reef would have looked in Carboniferous times.

Blistering heat of the desert

As 'Scotland' drifted northwards to Saharan latitudes, the lush ecosystems of the Carboniferous gave way to a harsher desert environment that persisted for much of the Permian and Triassic Periods. By this time, all the landmasses of the early world had coalesced to form one supercontinent, Pangaea. The land that was to become Scotland was located in a land-locked position close to the continental edge.

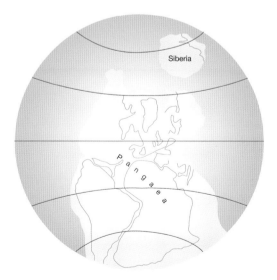

The British Isles lay close to the Equator during Permian and Triassic times, so the desert conditions preserved in the record of the rocks aren't unexpected. The rocks from this time indicate the presence of towering sand-dunes many tens of metres in height and the now fossilised remains of plants and animals consistent with this type of environment.

Model of *Elginia's* horned skull (left) and of the whole animal (right)

The climate was hot and dry, and ephemeral rivers criss-crossed the land. Huge sand dunes drifted across the parched landscape, driven by withering desert winds. It was a challenging world in which to survive. However, the sandstones of the Elgin and Lossiemouth area have yielded quite remarkable evidence of creatures that eked out a living in this hostile terrain. For example, small fragments from the reptile *Elginia mirabilis* were found near Elgin. The skull has 16 spines sticking out with the two largest looking like cow's horns.

The land was sparsely vegetated, but conifers, cycads and ferns were all present, growing around watering holes in the desert landscape. As the animals moved from place to place, they inevitably left tracks in the sand. Most tracks were lost under the

Sometimes there are no bones or other bodily remains for the fossil hunters to study. It may be just the tracks of a passing animal preserved in the wet sand that are left for us to ponder. But such remains give clues as to the size, weight and form of the animals that made them. Here are some reptile tracks at Clashach near Elgin. They were made by the reptile *Cheilichnus* as it walked across the surface of a sand dune.

This is the reptile *Stagonolepis*. The fossilised remains, which allowed this reconstruction to be made, were recovered from Late Triassic sediments near Lossiemouth. The reconstruction of the environment in which this creature lived is typical of the scene that would have existed across 'Scotland' just over 200 million years ago. From nose to tail, *Stagonolepis* was about 3 metres in length with short legs and a long tail. The teeth of this creature indicate it was mainly a vegetarian. Its 'armour plated' outer layer was probably just for defensive purposes. Meat-eating reptiles have also been found in nearby rocks of a similar age, so these protective layers would have come in very useful!

shifting dunes, but where conditions were right, probably in areas where the sand was slightly moist, footprints made by these early reptiles were occasionally preserved. Trackways found in Clashach quarry near Hopeman, Moray, are some of the best examples.

The Permian ended with another 'great dying' with 95 per cent of all marine and 70 per cent of all land-living species dying out. The cause this time was widespread volcanic eruptions that had the effect of poisoning the atmosphere and the oceans. Once again, life on Earth was hugely impoverished by natural causes.

Into the Triassic Period, conditions remained equally harsh and uncompromising. Despite the environmental upheavals created elsewhere in the world, life in the desert continued. Reptiles such as *Stagonolepis robertsoni*, a sleek creature with short legs and a long tail, grew to 3m in length. It was probably a vegetarian with its upturned snout being used to root out desert shrubs. The remains of another reptile, *Leptopleuron lacertinum*, have also been found in the Elgin area.

Shallow, tropical seas teem with life . . . including dinosaurs!

A more green and pleasant feel to the landscape was established in Jurassic times. Pangaea split asunder as yet another rearrangement of land and sea blocked out the outline of the continents that are familiar to us today. 'Scotland' waved goodbye to North America and Greenland, as the North Atlantic Ocean began to open during these cataclysmic upheavals. This movement shunted 'Scotland' further north to about 35° N of the Equator, maintaining its direction of travel out of the Tropics.

Left.
Reconstruction of the plant-eating dinosaur *Cetiosaurus*, whose remains have been recovered from rocks on the Isle of Skye.

Right.
Dinosaur expert Dr Neil Clark points to the footprint thought to have been made by a large carnivorous dinosaur. The sediments into which the large clawed foot was thrust some 160 million years ago also contain many useful indicators that help us to reconstruct the environment in which this creature lived. The ripple marks on the rock surface indicate the deposits were laid down in shallow water, possibly at the edge of a lagoon. The sand also contain numerous burrows made by worms and other burrowing creatures.

As every schoolchild knows, the Jurassic was the age of the dinosaurs. Across the globe, they occupied most of the ecological niches. From the *Pterodactyls* that patrolled the skies to the lumbering *Diplodocus* with its long neck and whip-lash tail, the dinosaur dynasty was a towering global force across the surface of the planet.

In Scotland, thick deposits of Jurassic strata on the Isle of Skye have yielded both plant- and meat-eating dinosaur species.

Evidence of dinosaurs takes the form of fossilised bones recovered from the rocks and also footprints that survive to this day in ancient sands. These trackways are powerful evidence for the presence of both meat- and plant-eating animals that inhabited the coastal plains of Jurassic Scotland some 150 million years ago.

During his voyage around the west coast of Scotland aboard the good ship *Betsey* in 1844 and 1845, the eminent Victorian palaeontologist Hugh Miller, who made his name collecting fossil fish in the Black Isle, made some amazing discoveries on the Isle of Eigg. Fossilised bones from what are thought to be early crocodiles were recovered from Jurassic deposits, along with vertebrae

This is a re-creation of the Jurassic seas as they may have existed some 165 million years ago. It was in this type of environment that the Jurassic rocks of Skye were deposited. The scene shows an ichthyosaur (a dolphin-like marine reptile), ammonites, belemnites and a shark scavenging the carcass of a dead dinosaur in the far distance.

and rib bones from a plesiosaur. Plesiosaurs were large marine reptiles that lived in the warm shallow seas of the time. They are only distantly related to dinosaurs and more closely allied in evolutionary terms to lizards and snakes. The plesiosaurs shared the oceans with ammonites of various types and a mollusc with a bullet-shaped shell – the belemnite.

Into Cretaceous times, global sea levels rose to unprecedented levels and 'Scotland' was drowned under the advancing waves. Sea levels stood at an amazing 300m higher than they are today

Left.
The ammonite *Pictonia* from a collection curated by the National Museums of Scotland. This specimen was collected from Shandwick, Ross-shire.

Right.
Female (large) and male versions of the ammonite *Ludwigia*. Both these fossils were recovered from Berreraig Bay on Skye.

Left.
A beautifully preserved leaf from the Ardtun on the Isle of Mull. It is thought that much of this part of the world was covered by forest around 55 million years ago, but periodic volcanic eruptions scarified the surface, turning a verdant scene into one of utter desolation. The varied flora of plane trees, ferns, cycads and conifers is preserved in ancient soils found between successive lava flows. There was clearly time for colonisation of the cooled lava surface to take place before the next eruption of molten lava swept across the land.

Right.
MacCulloch's tree on the Isle of Mull. It was estimated to be just short of two metres in diameter and around 16m in height. See facing page for a further explanation.

and all but the highest ground was submerged. What is even stranger about this period is the lack of any significant imprint left on the land. This was the age of the chalk seas of southern England, and the white cliffs of Dover, prominent landscape features today, were bequeathed from these times. But in Scotland, just a few remnants survive of this flood of biblical proportions. The strata of this age in Scotland have yielded an unremarkable shallow-water fauna that included starfish and marine sponges.

The end of the Cretaceous Period saw the fifth 'age of dying', when the dinosaur dynasty ended. The most likely explanation for this mass extinction is that about 65 million years ago a meteor struck the Yucatan Peninsula in Mexico. This set in motion a chain of events that led to global environmental changes, wiping out the dinosaurs in a geological instant. Clouds of dust thrown into the atmosphere by the force of the impact blocked out the Sun for years. Tsunamis swept across the land, and forest fires raged. All but the most resilient perished.

By Palaeogene times, a benign sub-tropical climate prevailed. 'Scotland' was a turbulent place of many active volcanoes. Between eruptions, layers of soil developed that supported a rich

flora. Broad-leaved deciduous trees co-existed with conifers, ferns and *Ginkgos*. The most prolific fossil site is Ardtun on the Isle of Mull, where an ancient soil, sandwiched between successive lava flows, is rich with plant remains.

An amazing discovery was made by John MacCulloch, itinerant geologist and celebrated map-maker. On his travels across Mull in the early part of the nineteenth century, he chanced upon the erect trunk of a conifer tree entirely encased in basalt lava. Little remains of this mega-fossil today – erosion by the sea and fossil collectors have seen to that. But his reliable description and drawings are convincing evidence of the existence of a mature stand of redwood-type trees that were subsequently invaded by a lava flow erupted from the nearby Mull volcano. There are only a very limited number of places where we get tantalising glimpses of this woodland ecosystem, but it seems reasonable to assume that much of the country was widely forested between the many volcanic episodes that scarified the early landscape.

Getting chilly? The Ice Age looms

The balmy climate of the Palaeogene gave way to something altogether more extreme during the final period of our journey through time. The Earth was now in the thrall of an ice age. During the last 2.6 million years, the climate has fluctuated wildly, with periods of deep freeze punctuated by times of relative warmth. The flora and fauna adapted to meet the challenges of these extremes of climate and weather. We know most about the final glaciation, as each re-advance of the ice largely obliterated the evidence of the previous one.

This period also saw the appearance of *Homo sapiens* as a dominant force in the ecosystem that emerged from the final retreat of the ice around 11,700 years ago. Although this is still the subject of intensive research, the accepted wisdom is that our species developed in the valleys and plains of East Africa and, over a prolonged period, migrated northwards and eastwards to populate Europe, Asia and then the rest of the world. Early humans reached northern Europe by around 700,000 years ago and would also have populated southern Britain by around that time. The English Channel was no barrier because, at that time, it didn't exist! Britain was just a northern extension of mainland Europe.

When the Ice Age was drawing to a close, some of the water on Planet Earth was still locked up in the ice sheets that covered much of Northern Europe and similar latitudes elsewhere. As a direct consequence, sea levels were considerably lower than they are today. This revealed an area of land that linked Scotland with continental Europe. Scientists called this Doggerland. As the climate warmed, the ice melted and sea levels rose. Doggerland now lies at the bottom of the North Sea.

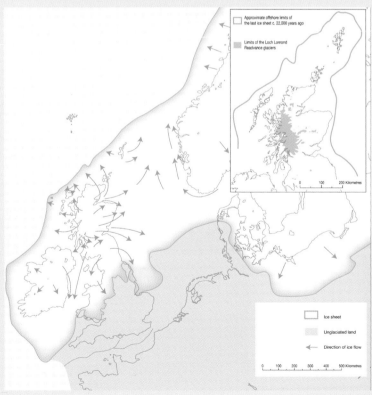

The ice extended from the North Pole southwards across the North Sea to cover much of Britain. For many thousands of years, Scotland was swathed in a thick blanket of ice. The photo on the facing page is from Antarctica today and there would have been a similar scene across the Highlands of Scotland for much of the Ice Age. It was at this time that the mountains and glens, straths and rolling hills of Scotland were shaped. The thickest accumulations of ice were to be found in the centre of the country and tongues of ice, known as glaciers, moved outwards propelled by their own weight in the direction shown by the arrows. As they ground across the land, so the scenery that we now recognise was created.

After the end of the last glaciation, much of the southern part of the North Sea was also dry land, known as Doggerland. It covered an area of similar size to that of modern-day Germany, linking Britain to the continent in one uninterrupted swathe of low-lying land. During the colder periods of the Ice Age, much of the water on the face of the planet was locked up in ice and snow fields, so, worldwide, sea levels plunged to much lower levels than they are today. Doggerland was thought to be home to around 100,000 people in its heyday. But, as a result of the rise in sea levels due to rising temperatures melting the world's ice sheets, that rich cultural landscape is now at the bottom of the North Sea.

The final glacial advance buried Scotland in a continuous blanket of ice to a depth that probably covered all but the highest summits. The landscape then was similar to that of Antarctica today – a great expanse of ice and snow with occasional peaks of jagged rock breaking the surface. At this time and other periods of glacial maxima, the sea levels were commensurately low with

Antarctica today: what Scotland would have looked like during the last glaciation

During the last glacial episode, ice covered the Cairngorms between 30,000 and 15,000 years ago. At the base of this thick icy blanket, fast-moving streams of ice cut particularly deep into the bedrock. The Lairig Ghru in the Cairngorms is a good example of this phenomenon where the ice has sliced a deep channel through the heart of the granite massif. More recent landslides down the sides of the glen have softened the landscape somewhat.

snow and ice extending from the poles.

But as the climate warmed the ice melted, revealing a deeply scarred and barren landscape. Glaciers and fast-moving ice streams within the ice sheet had cut deep into the fabric of the land. Great glacial gouges were revealed and piles of debris dumped by the ice, known as moraine, littered the landscape. It was a raw and inhospitable place.

In this hostile world, a few pioneer plant species clothed the land. Juniper and birch scrub were first to arrive, followed by alder, pine and oak. A hugely diverse population of mammals also appear in the fossil record. One of the best places to find such evidence is within the layers of sediments that build up within cave systems, which then lie largely undisturbed by later interference. One of the most intensively studied cave systems is Creag

The Creag nan Uamh cave system near Inchnadamph was home to a wide variety of species, including our ancestors.

nan Uamh in the Northwest Highlands. It was first excavated in 1889 and has been the subject of research ever since. The accumulated list of animal species recorded from this one site is truly remarkable and spans the time from the last glaciation into the postglacial period. Brown bear, Arctic fox, northern lynx, wolf, reindeer, Arctic lemming, water vole, common shrew, wildcat, stoat, badger, pig, red deer, sheep, ox, rabbit, hare and finally us – humankind. At another location in southwest Scotland, the remains of a mammoth have been found, so the cold-climate signature from these sites is definitely apparent.

Our appearance in the fossil record was presumably due to using the cave as a shelter from the elements. Later, our ancestors constructed their own structures. At Skara Brae in Orkney, the remains of a cluster of houses constructed in Mesolithic times around 4,500 years ago was uncovered during a winter storm over a century ago. Now a World Heritage site, this place marks the emergence of humans in Scotland living an existence we can recognise. So our story must end here, as we are beginning to encroach upon the archaeologist's and historian's jealously guarded territory!

It is interesting to reflect that if the history of life on Earth from earliest times was equivalent to a book running to many thousands of pages, humans would make an appearance in that story only in the final sentence. So although we are now the dominant ecological force on the planet, humankind is merely a bit-part player in the context of the preceding pageant of life on Earth.

The 4,500-year-old furniture of the Mesolithic people who inhabited this house at Skara Brae is remarkably like our own! But without a ready supply of wood, the material of choice to construct the cupboards and sleeping areas is local stone – Orcadian flags.

CHAPTER 5

A geological legacy

Scotland today is hugely influenced by its geological past. The location of mountain ranges, lochs and other iconic landscape features, the distribution of land suitable for growing crops and the places where great towns and cities were founded are matters largely pre-determined by its rocky heritage. Even the appearance of settlements is influenced by the building materials available to the architects, which in turn was determined by local geology. Compare, for example, the classical architecture of Edinburgh, Scotland's Athens of the North, with the grey granite buildings of Aberdeen. The difference in style is largely determined by what lies beneath our feet.

The wider landscape is also a product of the past. The Highlands were created when the Iapetus Ocean closed some 420 million years ago and have remained high ground since that time. The rocks produced by these geological processes are tough and have resisted erosion by the elements. In contrast, the softer, more douce countryside of central Scotland and parts of the Scottish Borders result from the bedrock having been worn down more easily by wind, water and ice.

When the Industrial Revolution took hold in Scotland, new mining settlements were located in Fife, between Edinburgh and Glasgow and southwards into Lanarkshire and Ayrshire because the exploitable coal resources that powered these social changes were located in these areas. Although the coal mining industry in Scotland has been consigned to history, these areas remain some of the most heavily populated parts of the country to this day.

Road and rail transportation routes which link communities followed the lead taken by the glaciers. The ice carved great pathways through the higher ground that engineers would be hard pressed to emulate. So they did the sensible thing and took advantage of the work already done by the glaciers.

The acclaimed elegance of Edinburgh's New Town is in part due to the fine building stone locally available to the architects and builders.

Aberdeen is a city built from granite. Marischal College is in many ways the centrepiece of the city's architecture. The poet John Betjeman commented on the college and adjacent buildings thus: 'Bigger than any cathedral, tower on tower, forest of pinnacles, a group of palatial buildings rivalled only by the Houses of Parliament.'

The Rest and Be Thankful has been a thoroughfare through the Argyllshire Hills since a military road was constructed here in the 1740s. It was much earlier than that when the ice cut a swathe through these uplands. Since that time, it has afforded an easy passage for travellers from Loch Lomond westwards to Inveraray and the open sea at Loch Fyne.

LEISURE AND PLEASURE

The championship golf course at Gleneagles was formerly a glacial landscape. The ridges either side of the mown grass are eskers that formed as the ice melted. They now afford an excellent viewing platform for spectators at this famous golfing venue.

Even where we take our leisure is determined by times past. Gleneagles, in rural Perthshire, was the venue for the Ryder Cup in 2014. It is one of the best championship golf courses to see all the action, because its architecture is a result of what happened at the end of the Ice Age as the ice melted. Great torrents of meltwater coursed under the melting ice sheet, charting a twisting path across the landscape. Like any other river in spate, these rivers carried boulders, sands, gravels and mud eroded from the glacier. As the gradient slackened, the sediment load was dumped. Constrained by walls of ice on the underside of the glacier, a new landform with steep sides was created – the esker ridge. Gleneagles is criss-crossed by eskers that were later utilised by the master golf course architect James Braid. Where parallel esker ridges exist at Gleneagles, the fairways snake between them. The higher ground that the eskers provide is the perfect viewing platform for spectators.

Hill walking and rock climbing are also hugely popular pursuits, and Scotland's hills and mountain ranges provide a cornucopia of possibilities. The 284 summits in Scotland that stand over 3,000 feet were first listed by Sir Hugh Munro of Lindertis in 1891. The publication of this 'Table of Heights' gave rise to what was to become, for some, an obsessive desire to conquer all the peaks on Sir Hugh's list – an activity that became known as 'Munro-bagging'. Almost all the mountains on the list

A GEOLOGICAL LEGACY

Sir Hugh Munro's list of the highest peaks in Scotland largely comprises mountains made from the toughest of rock – usually igneous or metamorphic in composition. But Liathach, which forms the northern flank of Glen Torridon is fashioned from Torridonian sandstone.

are made of the toughest stuff – igneous and metamorphic rocks that are found largely in the north and west of the country. Again, geology calls the shots.

But with every rule come a few exceptions. Some of the sandstone peaks of the Torridon Hills and elsewhere in Wester Ross also qualified for Sir Hugh's list.

Central and southern Scotland are entirely unrepresented because the softer rocks here succumbed to the erosive action of the ice and were moulded into altogether less impressive summits.

Some of the highest Munros create the peaks and pitches

The Cuillin on Skye is one of the most popular locations in Scotland for serious climbers to test their skills.

beloved of mountaineers. The granular gabbros of the Black Cuillin of Skye provide ideal footing for the adventurers who have tackled the line of peaks forming the Cuillin Ridge from Sgùrr nan Gillean in the north to Sgùrr nan Eag three miles to the south. Some even dare to climb all 11 Munros in the Skye Cuillin in one expedition.

SCOTLAND'S SOILS

On the jigsaw of assorted rock types of different ages, as described in earlier chapters, a rich and diverse community of plants and animals has developed since the ice melted. But, without a growing medium, none of this burgeoning biodiversity would have been possible. So let's talk about soils. They are perhaps something we take for granted. For most people, soils are just somewhere to plant our potatoes and dahlias, a comprehensively dull and uninteresting dirty brown layer of stones, mud and sand. But if soils didn't exist, almost nothing would grow, and the surface of our planet would be little more than a lifeless rocky place.

By the time the last glacier melted around 11,700 years ago, the landscape was covered by a generous, chaotic and non-

As ice cover across Scotland melted, so the burden of rocks and other debris carried by the ice – mud, sand and larger boulders – was dumped.

uniform layer of glacial debris. Scottish soils largely developed from this parent material bequeathed from the Ice Age.

The exception is where the cover of dumped glacial debris was thin or indeed non-existent. In those circumstances, the nature of the bedrock is the main determinant of the type of soil that subsequently developed.

The process of soil development started as soon as the ice melted away, and it continues to this day, influenced by many factors such as altitude and climate, lie of the land, movement of water and the speed with which the area was initially covered by pioneer species of trees and small shrubs.

Our soils have only been developing since the end of the Ice Age, so are considered to be 'young'. In areas of the world where soil development has continued uninterrupted by ice ages or other upheavals, soil formation has been going on for millions of years, resulting in some instances in soil profiles that are many metres deep.

Soils are home to a myriad of micro- and macro-organisms. Soils have been compared with the rainforests of the Tropics in terms of the number of species of plants and animals they support. Just one teaspoonful of soil may contain up to several million micro-organisms – bacteria, fungi and protozoa, in

Soil development is limited across many of the upland areas of Scotland. The soils are characteristically thin and are able only to support heather and other hardy plant species.

In many lowland areas of Scotland, fertile and well-managed soils form the basis of one of our most successful industries – arable farming.

addition to a selection of algae, microscopic worms called nematodes, beetles, mites, caterpillars, ants and the occasional mole. Many of these species play a vital role in the way that soils function, including decomposition of organic matter and the recycling of nutrients.

Soils carry out many functions that help us to live a comfortable life. They act as a growing medium for our crops; they absorb rainfall and provide a buffering role, so that the land doesn't flood every time it rains. Peat, in particular, fulfil a vital role in acting as a carbon store – especially important in these times when minimising our carbon footprint is a key issue. Soils also make the countryside what it is, supporting the wide variety of habitats for which Scotland is well-renowned.

SCOTLAND'S NATURE

Soils are hugely important in determining which habitats can be supported, so heavily influence the look of the countryside. For example, deciduous forests grow on brown earth soils; coniferous forests and heathlands on podzols; and rushy pasturelands are mostly associated with gley soils.

In determining 'what grows where' and therefore the distribution of habitats throughout Scotland, there are many factors at work.

Climate is key. There are significant differences in rainfall, wind speed, humidity and temperature across Scotland, and these are important factors in determining the type of vegetation cover that thrives and survives in each part of the country.

Altitude also plays an important part. There is a marked vegetation change from the foot to the summit of mountain ranges. In this 'environmental gradient', perhaps the most important factor is the location of the tree-line. Above the tree-line, the countryside is dominated by hardy shrubs, grasses and other low-growing plants. This is the sub-alpine zone.

The drainage characteristics of the soils are another important determinant of vegetation cover. Where drainage is poor, bog-moss or *Sphagnum* is the predominant vegetation that develops. It is one of the most widespread habitats in Scotland with around one million hectares covered by peatland.

Let's consider two of these habitats, fashioned by geological processes, in more detail:

Deciduous woodlands, such as this attractive bluebell wood, are commonly found in lowland areas of Scotland. The characteristic growing medium for this type of habitat is a brown forest soil.

Coniferous woodland are found in many parts of Scotland. A creation of the Forestry Commission, these regimented plantations grow in the most widespread soil type in Scotland – the podzol.

Heathlands cover a high proportion of Scotland. The soils of this upland environment are thin and acid in nature. They are also most likely to be classified as podzols.

Raised bog of Flanders Moss

Flanders Moss National Nature Reserve in the Forth Valley is an excellent example of a raised bog. Think of a national nature reserve and you are unlikely to picture a giant compost heap. But that is exactly how Flanders Moss could be described! It is a relic from a bygone age, dating back 11,700 years and more to the end of the Ice Age. As the ice melted, a lunar landscape of dumped glacial till and assorted debris emerged from the blanket cover of ice. In areas where the drainage was poor, a boggy wilderness developed and layer upon layer of vegetation accumulated to form a raised bog. It is called 'raised bog' because it has grown to a level higher than the surrounding land. At one time, these features were commonplace in the central part of Scotland, but agricultural improvements and drainage works over the past few centuries have destroyed many of these raised bogs. Those that remain have become all the more precious.

Flanders Moss also holds evidence of huge environmental changes in the layers of peat that created the structure. Pollen grains from plants that grew close by or indeed formed part of the bog became trapped in the peat as the layers built up over millennia. As a rough guide, a metre of peat takes around a thousand years to accumulate. Sampling the surviving pollen grains from various levels within the peat bog gives us a picture of changing patterns of vegetation and other events that affected the area.

Flanders Moss National Nature Reserve is home to a wide variety of plants and animals, but its main claim to fame, in a geological context, is its link with the past. It is a habitat that developed in the immediate aftermath of the Ice Age, untouched and undrained by the 'agricultural improvements' that destroyed many similar features across the central belt of Scotland.

The first plants to establish in these desolate conditions in central Scotland were juniper and birch scrub, followed by an extensive cover of oak, hazel and elm. Human influences can also be picked up in the fossil record. There are marked declines in elm pollen around 5,800 years ago, coinciding with forest clearances during the Neolithic and Bronze Ages. Ancient remains, such as a Bronze Age bucket, swords and Britain's oldest wheel, show that this part of Scotland has been inhabited for thousands of years.

Wildlife is abundant on the moss. Adders and mountain hares are two of the creatures that thrive on the bog all year round, while thousands of migrant pink-footed geese from Iceland and the Arctic roost on the lochans over winter. Cranberry, bog rosemary and bog myrtle are amongst a fairly small assemblage of plants found here. Other notables include a rare species of jumping spider, carnivorous sundews that use sticky hairs to traps insects on their leaves, and a host of dragonflies and butterflies.

The machair of the islands and Northwest Highlands

Machair is an iconic Scottish landscape created by the natural processes of wind and waves. This habitat is most widespread on the western coastline of the Western Isles, where it runs in an unbroken ribbon for almost 110 kilometres. It is also found on the islands of Coll, Tiree, Colonsay, Mull and also Orkney and

Seilebost beach on South Harris is a fine example of a beach made from broken shells. The machair lies adjacent to these shell sand beaches in parts of Lewis and Harris and is also prevalent in the west coast of the Uists at the southern end of the Outer Hebrides archipelago.

The machair in full bloom is an amazing carpet of flowers.

Shetland. A few outposts also decorate the coast of the Northwest Highlands. Machair takes its name from the Gaelic word meaning low-lying fertile plain and is the basis for one of the oldest ways of life in Scotland – crofting.

This fertile landscape is built from broken shells and glacial debris dumped offshore towards the end of the Ice Age. Many shelled sea creatures lived and died along the length of this coastline. Under the influence of crashing waves, the shells were broken and the debris swept ashore by the action of tides and currents. The mass of broken shells mixed with the glacially derived sands was then blown inland by the prevailing winds to create coastal dunes and fertile pastures beyond. The beaches created from this material are a dazzling white colour.

The sand plain inland from the coastal dunes has been cultivated by crofters for generations. Oats and rye are grown in rotation with potatoes. During the winter months, the land is grazed by cattle.

This way of life is often supplemented by an income from another source, such as employment as the local postman or provision of accommodation for tourists. But this fragile existence may be under threat from nature. Rising sea levels and lack of natural replenishment of beach material from offshore mean that

One of the rarest species of bumblebee in the UK, *Bombus distinguendus*.

the machair may, in future decades, fragment and diminish in size and extent. But that is for the future. We should appreciate this unique habitat while it is still with us.

The shell sand of the machair provides a home for some of Britain's rarest creatures, such as the bumblebee *Bombus distinguendus*. This species lives in old rodent burrows that have already been excavated into the soft sand.

A DYNAMIC LANDSCAPE

The ice may have left the scene and the countryside seems unchanging, but the landscape continues to evolve. Slopes appear stable, particularly during the summer months, but after periods of prolonged rain, they are prone to collapse without warning in a catastrophic manner that threatens life and limb. We stop to admire and photograph the dramatic slopes carved by the passage of ice, but the thin mantle of glacial till that smears the valley sides is prone to slope failure. In 2004, after a period of heavy and prolonged rainfall, a major landslip occurred in Glen Ogle at the eastern extremity of the Loch Lomond and Trossachs National Park. In this, the most dramatic landslip to occur in Scotland that year, 57 people were trapped in their vehicles and

The aftermath of the 2004 landslip in Glen Ogle. The rescue is in full swing with helicopters ferrying the stranded motorists to safety.

had to be airlifted to safety after two major debris flows swept down the mountain-side, wreaking havoc in their wake. There were no injuries on that occasion, but the landslip is a clear illustration of the fact that these passes through the mountainous landscapes of upland Scotland must be treated with respect.

Weather and climate are key factors in triggering such events, but, despite considerable research, such episodes of instability are difficult, if not impossible, to predict with any degree of accuracy. So the precautionary principle is adopted, and many slopes that were considered to be prone to movement have now been stabilised. Although public safety considerations must always be paramount, it is unfortunate when the 'naturalness' of the landscape is thus eroded.

A GEOLOGICAL LEGACY

Power station chimneys have belched out greenhouse gases – not just here in Scotland, but across the world – for decades. As most now acknowledge, they cause long-term changes in the climate of the planet.

CHANGE IS A CONSTANT

There are no habitats remaining in Scotland that can be described as pristine and in a fully 'natural' condition. Scottish wilderness is long gone. Human activity, in the form of farming, forestry or engineering works, has altered the natural condition of the landscape over the centuries.

Human reach extends from the coastal areas to the tops of the Munros through management of sheep and deer; planting of single-species woodlands; intensive farming of the lowland areas; and the construction of hard infrastructure such as towns, cities, roads, railways, power lines and most recently windfarms. This is not a conservationist's rant. There is much to be admired in the way we have fashioned the world around us to be a series of attractive, productive and varied landscapes. But let's not fool ourselves that much of what we see in the Scottish countryside today is entirely as nature intended.

Since the Industrial Revolution started over 250 years ago, we have been adding greenhouse gases to the atmosphere at an increasing rate and the effects of these actions are beginning to be felt.

There has never been a time in our geological past where things have stayed the same for long, so change has been a constant throughout history. However, we are now speeding up the rate of change. The human animal has been a significant part of the ecosystem in Scotland for the last four millennia and human actions have a global reach. Despite political interventions to slow down the rate of greenhouse gas emissions on an international scale, nothing of significance has changed or is likely to in the foreseeable future. We are hopelessly addicted to the use of energy to create comfortable homes, productive offices and factories, and efficient transport systems. Much of the energy required to power this lifestyle is currently supplied by fossil fuels.

Massive new reserves of coal are being exploited in China, Australia and other parts of the world, so this dependence isn't going to lessen anytime soon. Some 1,000 billion tonnes of coal remain in the ground worldwide and, at the current rate of use, there are several centuries of supply left to extract. Similarly, the first oil well was drilled in Pennsylvania in 1859, and reserves of

Melting ice caps will add many millions of extra gallons of water to the world's oceans. The inevitable consequence is that sea levels will rise. However, the melting of floating sea-ice, such as currently exists at the North Pole, creates no overall increase in sea level.

oil and gas will last for many decades to come. Together fossil fuels currently fulfil 80 per cent of the world's energy needs. Although the exploitation of renewable energy sources, such as wave, wind and solar, is welcome, it will be many years before it makes a significant global contribution.

Sea levels will rise as the land-based ice reservoirs at the South Pole and northern latitudes generally continue to degrade. Thermal expansion of the world's oceans will also contribute to the process of sea level rise as temperatures continue to increase.

There is hard evidence that sea levels have been up to 30m higher than present in the recent geological past (albeit thousands of years ago), so it should come as no surprise if we envisage significant global sea-level rises in the future. Every country in the world will be affected. Areas of coastal land where the sea currently benignly laps against the shore will disappear under a rising tide.

This is not a guess or a wild projection. Doggerland – that expanse of land the size of Germany that lay between Britain and continental Europe – was consumed around 8,000 years ago when sea levels rose in response to the melting ice caps.

Some might describe this as a doomsday scenario or a flight of fancy, but for others it is just a new reality to which we must adapt. Planet Earth has dealt with change since its formation some 4.54 billion years ago and inevitably will do so again. So there is no need to worry about saving the planet, as it is endlessly resilient and adaptable.

On the Isle of Arran, cliffs and stacks carved by the sea were left high and dry as the high-tide line retreated by 100 metres or more. The intervening stretch of land, now grazed by sheep, is known as a raised beach and would have been the inter-tidal area when the sea level was higher. The island was pushed down by the weight of ice during the last Ice Age, but 'bounced back' once the burden was lifted as the ice melted. So the land rose relative to the sea.

Above.
'Save the Planet' protesters slightly miss the point! Planet Earth will sail on for a few billion years more regardless of what environmental insults we throw at it. What is really at issue is humankind's ability to continue living on the Earth in a long-term and sustainable fashion.

Below.
Heavy rain leading to flooding (in this instance on the River Tay) and other extremes of weather are predicted to be more frequent if the climate change scenarios currently envisaged prove to be correct.

Our climate is also becoming more prone to extreme events. So when it rains, it rains more heavily and for longer. Extreme droughts and much higher winds are also becoming more prevalent, ensuring our world is becoming a 'stormier' place.

The key question for humankind is: how will we adapt to a different climate and an unfamiliar geography? We urgently need to understand which changes are most likely to happen and the rate at which these modifications to our world are likely to take place. Only then can we plan to accommodate the different landscape that will inevitably await future generations. Some countries will lose a little of their current territory; others will lose a lot. Significant change may not materialise for a generation or two, but if we continue to pump greenhouse gases into the atmosphere at the current rate, there will only be one outcome. The only significant matter about which the jury is still deliberating is the extent to which sea levels will rise and the timescale over which these changes will happen.

WHERE NEXT FOR THE SCOTTISH LANDSCAPE?

All that we can say with absolute certainty is that the Scottish landscape of the future will look very different from that of today. In the short term, this green and pleasant land will continue to heat up as temperatures rise in response to global warming. Scottish landscapes and the biodiversity they support are already showing demonstrable signs of change. An example documented

by Scottish Natural Heritage is that plants previously well-adapted to grow in the alpine zone of the Cairngorm plateau are now disappearing from the area as temperatures rise. Distribution of birds and insects is also changing as more species previously associated with warmer southern climes are now routinely found in Scotland.

Longer-term environmental change is also afoot. We are currently in an interglacial period where the ice has melted and the climate has warmed. This has allowed life to flourish over the last 10,000 years. But, despite all the reports of global warming and melting ice, we are eventually destined for another global ice age. The timescale for these changes is many tens of thousands of years hence. Some estimates put the onset of the next glacial episode at around 50,000 years from now, taking into account likely carbon dioxide emissions added to the atmosphere during this period. The onset of the next ice age would have been rather sooner without this contribution.

Another great unknown is the impact of one-off events, such as the eruption of a super-volcano. They are like googlies bowled by Nature! The slumbering giant that lies beneath Yellowstone Park in the USA has been dormant for thousands of years. In fact, it last erupted 640,000 years ago, when it is estimated that 1,000 cubic kilometres of rock and volcanic ash were ejected. When it does erupt in the future, as it inevitably will, the effects will be felt worldwide. Once again, ash and rock will be hurled high into the air and the rays of the sun will be blocked out. This will have a considerable effect on temperatures and air quality around the globe. Every part of the world will feel the effects of this event. But no-one can predict with any degree of accuracy when the 'big one' will happen next.

THE ONSET OF THE NEXT ICE AGE

We have been in an ice age for over 2 million years. Many advances and retreats of the glaciers have already scoured and sculpted the surface features of Scotland during that period. Recently, we have basked in benign conditions since the last ice melted around 11,700 years ago: we're in an interglacial period. As the graph on p. 90 shows, there have been many warm periods in the recent geological past, but the ice inevitably returns each time.

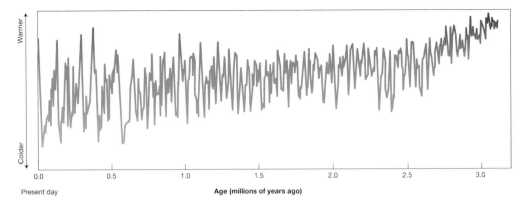

Over the last few million years, global temperatures have peaked and troughed. The troughs are characterised by glaciations or re-advances of ice from the poles whereas the peaks are warm periods or inter-glacials. We are currently experiencing a warm period, but the ice is expected to reach down from the North Pole once again smothering Northern Europe in a blanket of snow and ice, perhaps in another 50,000 years or so.

There are three main factors that determine the timing of the onset of the next Ice Age:

- Planet Earth's orbit around the Sun varies from elliptical to circular over a cycle that lasts between 90,000 and 120,000 years. At certain times of the cycle, the planet is further away from the Sun meaning that colder conditions ensue. (see A. Eccentricity, on facing page)
- The axis around which the Earth rotates is not vertical. This brings the northern and then southern hemispheres closer to the Sun alternately during the Earth's annual rotation, giving rise to the seasons. The angle of tilt from the vertical varies over a 41,000-year cycle. (see B. Obliquity, on facing page)
- And, finally, the planet also wobbles on its axis as it achieves its daily rotation, just like a spinning top when it slows down. This wobble (or precession) has a 22,000-year cycle, varying from a small to a more pronounced eccentricity. (see C. Precession, on facing page)

Taken together, these factors affect the amount of heat from the Sun that reaches the surface of the Earth. Over long periods, these cycles combine to force long-term climatic changes, swinging the climate from pleasant and benign conditions to a renewed onset of a global winter that may last for many thousands of years. Our world will stand on the threshold of just such a change in around 50,000 years' time.

So change is locked into the way in which our planet functions, and in the future we have no choice but to adapt to these extremes of nature.

Opposite.
Wobbles and orbital cycles

A. Eccentricity

Elliptical

Almost circular

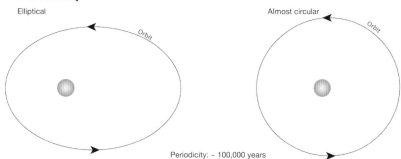

Periodicity: ~ 100,000 years

B. Obliquity

Tilt of axis

Periodicity: ~ 41,000 years

C. Precession

Wobble

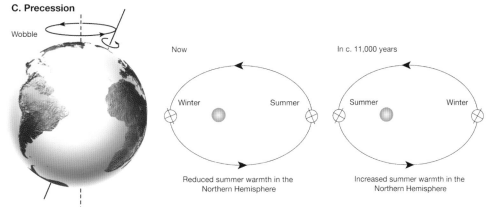

Now

Winter   Summer

Reduced summer warmth in the Northern Hemisphere

In c. 11,000 years

Summer   Winter

Increased summer warmth in the Northern Hemisphere

Periodicity: ~ 22,000 years

. . . AND FINALLY!

That concludes our brief sojourn through time from the earliest days of Planet Earth to the present time and beyond. We've seen how Scotland has developed during successive geological periods, as a builder might add an extension and then another floor to a much-loved house. On this journey through time, Scotland has moved from the southern hemisphere through all the climatic conditions the planet has to offer, colliding with other continents and then separating to create new worlds. Aboard this chunk of rock that was to become Scotland has been a changing cargo of plants and animals. From the primitive graptolites that inhabited the Iapetus Ocean to the ferocious dinosaurs whose fossilised bones are found on Skye, the changing biodiversity we see in the record of the rocks in Scotland is truly breathtaking.

Throughout this journey, change has been the only constant. The perspective that the study of geology gives us is that what holds true today will not necessarily hold true in the future. There are global drivers for change at work over which we have absolutely no control. Continents continue to move, driven by the deepest-seated of processes, and those species that survive the coming 50,000 years will be plunged into the next ice age by a combination of planetary wobbles and orbit changes. In taking this long-term perspective, it's clear we are not entirely masters of our own destiny.

Our most important imperative must be to rediscover the art of survival and learn to adapt to this changing reality. If we don't, the human animal may be added to the long list of species that had their day in the sun, but ultimately lost the battle for survival.

Glossary

aeon an indefinite, but lengthy, geological period of time.

ammonite an extinct fossil mollusc with a coiled shell. This animal group was particularly abundant during Jurassic times.

arthropods a group of animals well represented in the fossil record, including the now extinct trilobites and eurypterids (fossil water scorpions), with crustaceans (crabs, lobsters and shrimps) representing the group today.

caldera a large crater created by the collapse of surface rocks into an active volcano. Calderas are usually formed during explosive volcanic activity. This phenomenon was first described in Glencoe and subsequently recognised in many other parts of the world.

Caledonian mountain building the event spanning the Ordovician and Silurian Periods when the mountains of Scotland were formed.

continental drift the idea that continents move across the Earth's surface. When the German scientist Alfred Wegener first suggested the idea in 1912, the mechanism by which the continents were able to move was not known. The concept of 'plate tectonics' provided the answer.

core the centre of Planet Earth. It is too deep to be seen or sampled directly, but much is known of its chemical make-up and behaviour by remote sensing. It is mainly composed of iron with some nickel and small amounts of other elements. The outer layers of the core continually churn and it is this motion that generates the Earth's magnetic field.

crust the rocky outermost layer of Planet Earth. The continents are, on average, about 30km thick, whilst the crust that lies under the planet's oceans is substantially thinner at around 7.5km thick. Even the deepest drilling operation would get nowhere near the base of the Earth's crust.

dyke igneous rock associated with volcanic activity that has a linear form. Dykes vary greatly in thickness, but many are around 2–3 metres wide. Some dykes can be followed for many kilometres across country, forming a distinctive landscape feature.

ecological niche an environment on land or in the water that allows a particular species to thrive.

ecosystem a community of interacting living organisms (plants, animals and micro-organisms) that are supported by non-living environmental components – air, water and minerals.

erosion a process of wearing away the land surface by water, wind and ice. Wave action at coastal locations and the effect of rivers and streams are very effective means of modifying the landscape. Wind is a particularly powerful force in desert environments where sand grains, carried by the wind, act like emery paper. Perhaps the effects of ice cause the most dramatic changes to the landscape. The ice sheets that covered Scotland during the last Ice Age were up to 2km thick. As they scraped and gouged their way across the landscape, deep glens were excavated, and mountains were moulded into their now familiar shapes.

fissure a linear crack in the Earth's crust that allows molten rock to pass upwards to the Earth's surface.

fossils remains of past life, from both the plant and animal kingdoms, that existed in earlier times.

gastropods a group of animals dating from the Cambrian period, whose descendants live on to this day in myriad different forms – a snail is a typical example. In the fossil record, only the remains of the coiled shells are normally preserved, but, when living, the soft-bodied occupant of the shell would have possessed a head and foot.

geological column the device by which geologists have represented geological time from the very oldest events that occurred on Planet Earth to the youngest (see pp. 14–16). It includes the time when fossils where abundant, known as the Phanerozoic, from the Cambrian Period onwards to the present day – a timespan of around 545 million years; plus the much greater time span covered by the 'Precambrian', which stretches back to the earliest days of the planet.

geology the study of the Earth, its origins, structure, composition and life that the planet has supported since its formation. The Scottish scientist Dr James Hutton is widely acknowledged as being the founder of the modern science of geology.

gley gley soils are typically found in waterlogged situations where drainage is poor.

greenhouse gases gases such as carbon dioxide and methane, which collect in the Earth's atmosphere, causing reflected heat from the planet's surface to be retained. So the temperature of the Earth's surface rises – a phenomenon known as global warming.

habitat an environment (e.g. coastal, mountain, urban) where a characteristic community of plants and animals live.

Iapetus Ocean a long-disappeared ocean that once separated 'Scotland' from 'England'. Over time, it filled with sediments and, as the ocean closed, mountains were formed when the continents collided.

ice age an extended period when the climate is such that glaciers and ice caps cover much of the Earth's surface. There have been many ices ages during Earth history. The most recent ice age took place during the Quaternary Period. It had a profound effect in shaping the landscape into the familiar mountains and glens recognised today. The ice melted 11,700 years ago and we are now in an interglacial period with a relatively warmer climate. But the ice will return in perhaps about 50,000 years' time when the world will, once again, be plunged into a prolonged glacial winter.

igneous rocks [**basalt, gabbro, granite, volcanic ash, volcanic bomb, obsidian, pyroclastic flows**] rocks related to volcanoes or volcanic activity of one type or another. What distinguishes one rock type from another is colour, mineral

GLOSSARY

composition and size of the individual mineral grains that make up the rock. Basalts and gabbros are known as 'basic', because of the base-rich (calcium, magnesium) minerals they contain. Granites and related rocks, such as felsites and syenites, are known as 'acid' because they contain high levels of silica. Igneous rocks are commonly found in many parts of Scotland.

interglacial a warmer period between ice ages when the ice melts and sea levels rise. There have been many such warmer interludes during the current ice age.

intrusions and extrusions terms used by geologists to describe the way in which volcanic rocks are 'added' to the geological record. When molten rock spills onto the surface from a volcano, it is described as 'extrusive', so all lavas fall into this category. When rocks form part of the subterranean plumbing of a volcano (its magma chamber or associated dykes and fissures for example), these rocks are described as 'igneous intrusions'.

magma molten rock that subsequently solidifies to form igneous rock. When magma is erupted directly onto the Earth's surface, it is known as lava. When a volcano is active, the molten rock it erupts is stored under the surface in a magma chamber.

mantle the layer that lies between the Earth's crust and core. The lower layers are in direct contact with the core and this causes the mantle to slowly circulate like syrup being heated in a saucepan. This convection process causes the tectonic plates making up the Earth's crust to move.

metamorphic rocks [gneiss, schist, slate, hornfel] rocks that have been altered by heat, pressure or both. The original characteristics of the rock (mineral composition and colour) are lost as the rock is transformed from what it was originally (sandstone, shale, granite or basalt, for example) to a new form. Deep burial in the Earth's crust and the proximity of hot molten rock are two of the most frequent ways in which rocks are transformed or metamorphosed.

minerals the building blocks of rocks, whether of igneous, sedimentary or metamorphic origin. Each rock type (basalt, limestone or mica schist, for example) is made up of a characteristic assemblage of minerals. It is on that basis of mineral content that rocks are distinguished from one another.

molluscs a group of animals with the defining characteristic of possessing a shell. They are well represented in the fossil record from earliest times and many forms exist today. These are animals without backbones (invertebrates) and they inhabited a wide variety of environments including freshwater, marine and land-based habitats.

moraine a jumble of sand, clay, boulders and pebbles picked up by glaciers as they ground across the landscape during the last ice age. This material was then deposited in characteristic landforms at some distance from where the boulders and pebbles originated.

orogeny the process of mountain building: when continents collided during past times, thick layers of sediment that had previously been deposited in the waters that separated the land masses were squeezed as if in a vice. These sediments were changed or metamorphosed in the process and thrown upwards to form mountains. As continents wandered the globe throughout geological time, many continental collisions are known to have occurred. Every mountain range anywhere in the world (Andes, Urals, Alps, Himalayas,

Pyrenees, Scottish Highlands to name a few) was formed by colliding continents during this process of mountain building.

period a sub-division of the geological column. The purpose of defining periods of geological time is to ensure that there is consistency in describing the age of rocks, fossils and geological events worldwide. Many of these geological periods, such as Cambrian, Ordovician, Silurian and Devonian, were defined in Britain, demonstrating the strong influence of British scientists when the foundations of the science were laid.

Phanerozoic a period of some 545 million years, from the start of the Cambrian Period to the present day, which is distinguished from earlier times by the fact that fossils with hard parts (e.g. shells) are commonly found in rocks from this period of geological time.

photosynthesis the process by which green plants transform light energy from the Sun into chemical energy. Water and carbon dioxide combine to form carbohydrates used as food. Oxygen is generated as a by-product of this process.

plate tectonics a theory that explains how the Earth's tectonic plates move and interact with each other. The idea that continents have moved throughout geological history has been around for over a hundred years. Plate tectonics provided the explanation for how this complex process works

plug a mass of volcanic rock (circular in plan) that cooled in the central feed pipe that links the magma chamber to the main vent of the volcano. Plugs only become visible if erosion has cut deep into the structure of the volcano to reveal this feature.

podzol the variety of soil that covers much of Scotland. The surface vegetation that podzols support is usually coniferous woodland or moorland.

rock types three types of rock make up the Earth's crust and are found worldwide – igneous, sedimentary and metamorphic. All three categories are well represented in Scotland. The distribution of rock types in any given area is best represented on a geological map – see page 16 for a geological map of Scotland.

sedimentary rocks [conglomerate, sandstone, limestone, shale] strata that are the eroded fragments of other rocks. They build up in a variety of surface environments, such as deserts, oceans, beaches, rivers and freshwater lakes. Accumulations of sedimentary rocks have occurred throughout geological time, and these types of rock are well represented in Scotland.

subduction a process in which one tectonic plate dives down underneath a neighbouring plate (see page 5). As tectonics plates move across the surface of the globe, they come into contact with each other. In some cases, the plates slide past one another, but some contacts are more destructive, as in a subduction zone.

tectonic plates pieces of the Earth's crust. The surface of the planet is divided into seven large plates and about twelve smaller plates. Each is rigid and in constant motion. As the plates move, they rub past each other. The frictional forces of this contact generate shoals of earthquakes. Tectonic plates are of two types: an oceanic plate that lies beneath the world's oceans or a much thicker continental plate.

trace fossil a burrow, trackway or footprint that is evidence of the past existence of a particular animal.

trilobites a long-extinct group of animals that lived in shallow seas from Cambrian times to the end of the Permian Period. Their fossilised remains are widely distributed in sedimentary rocks around the world.

tsunami a fast-moving and often destructive sea wave, usually generated by earthquake activity. Tsunamis occur at the present day, but the characteristic deposits they leave behind are also recognised in the geological record.

tundra a treeless habitat that is characteristic of a cold and inhospitable climate. After the ice melted, but before the climate had warmed, tundra was widespread across Scotland. The landscape at this time, around 10,000 years ago, would have looked similar to Siberia today.

ultrabasic a type of igneous rocks rich in minerals with a high iron and magnesium content. Many rocks in this category have exotic-sounding names such as peridotite, picrite, lherzolite and serpentinite.

unconformity a feature that sometimes develops from an interruption in the deposition of a sedimentary rock sequence (see page 7). In summary, where earth movements interrupt the deposition of a rock sequence, those layers may be raised above sea level and subject to erosion and tilting. At a future time these layers of rock may, once again, be submerged under water and younger sediments piled on top. The contact between the two sequences is called an unconformity and may represent many millions of years of geological time. James Hutton was the first to observe this phenomenon in the field and correctly interpret its significance.

vent an opening in a volcano where molten rock reaches the surface.

water column a vertical section through lake or sea water, demonstrating differences in the properties (oxygenation, levels of pollution, acidity, etc.) of the water at various levels.

Further reading

The Dating Game – One Man's Search for the Age of the Earth, C. Lewis (2002, reissued edn 2012), Cambridge University Press, Cambridge. [An entertaining account of Arthur Holmes' search for a method of accurately dating rocks and geological events.]

Death of an Ocean – A Geological Borders Ballad, E. Clarkson and B. Upton (2010), Dunedin Academic Press, Edinburgh. [A lucid introduction to a much neglected area of the country – the Scottish Borders.]

Earth – The Definitive Visual Guide, 2nd edn, ed. J.H. Luhr and J.E. Post (2013), Dorling Kindersley, London. [This is an excellent guide to the inner workings of Planet Earth. It is beautifully illustrated and written in an accessible style.]

Edinburgh Rock – The Geology of the Lothians, E. Clarkson and B. Upton (2006), Dunedin Academic Press, Edinburgh. [An excellent introduction to the geology of the area around Edinburgh.]

Geology and Landscapes of Scotland, 2nd edn, C. Gillen (2013), Dunedin Academic Press, Edinburgh. [An excellent text for the slightly more advanced student of geology – a logical progression from this book.]

Geology of Scotland, 4th edn, ed. N.H. Trewin (2003), Geological Society, London. [This is the only academic text on this further reading list. Although this text is now over ten years old, it is an excellent place to find more in-depth information about any aspects of Scotland's geology.]

The Great Tapestry of Scotland: The Making of a Masterpiece, S. Mansfield and A. Moffat (2013), Birlinn, Edinburgh. [The story behind the tapestry with colour illustrations of each of the individual panels.]

Hugh Miller – Stonemason, Geologist, Writer, M.A. Taylor (2007), NMS Enterprises Ltd, Edinburgh. [An enjoyable read about a man who did much to promote the study of fossils in Scotland.]

James Hutton – The Founder of Modern Geology, D.B. McIntyre and A.P. McKirdy (2012), NMS Enterprises Ltd, Edinburgh. [An account of the life and times of one of the most important historical figures in geology.]

Land of Mountain and Flood, A.P. McKirdy, J.E. Gordon and R. Crofts (2009), Birlinn, Edinburgh. [A readable and well-illustrated account of the geology and landforms of Scotland, suitable for the beginner.]

Prehistoric, D. Palmer and others (2009), Dorling Kindersley, London. [A beautifully illustrated and written account of life on Earth since earliest times.]

Scottish Fossils, N.H. Trewin (2013) Dunedin Academic Press, Edinburgh. [A fascinating account of the fossils found in Scotland.]

Volcanoes and the Making of Scotland, B. Upton (2004), Dunedin Academic Press, Edinburgh. [A very readable account of a key element of Scotland's geology.]

Index

Aberdeen, 72, 73
aeons, 7, 93
Ailsa Craig, 14, 31, 39, 46–7
akmonistion zangerli, 60
ammonites, 65, 93
Anthropocene period, 14
Archaen period, 16
Ardnamurchan Point, 44–5
Arran, 8, 14, 15, 31, 39, 46–7
arthropods, 54, 57, 93
Arthur's Seat, 32
Auchterarder, 32
Bailey, Sir Edward, 35, 37
Baird, James, 74
basalt, 33, 36, 38, 42, 43, 67, 94–5
basement rocks, 17, 24, 40, 44, 46, 95
Ben Nevis, 37
Big Bang, 49
Black Cuillin, 32
boulders, 2
British Geological Survey, 16
bumblebee *(bombus distinguendu)*, 83
Burgess Shales, 50
Cairngorms, 36, 70, 89
caldera, 37, 42, 45, 93
Caledonian mountain-building, 23, 25–6, 36, 93
Cambrian period, 15, 50, 52–3, 94, 96, 97
Canna, Isle of, 43
Carboniferous period, 13, 15, 52, 58, 59, 61
Clark, Dr Neil, 64
Clerk of Eldin, James, 8, 46, 47
Clough, C.T., 35
coal, 29, 58, 59, 72, 86, 96
conglomerate, 21, 43, 96
continental drift, 29, 52, 61, 93
core, 5, 38, 42, 44, 46, 93, 95
Cretaceous period, 15, 51, 52, 65–6

Crieff, 32
crust, 2, 4–5, 14, 16, 18–19, 22, 23, 25, 26, 27–8, 35–6, 38, 40, 44, 47, 48–9, 55, 93–6
Curie, Marie and Pierre, 12
Darwin, Charles, 10
Devonian period, 3, 13, 15, 51, 52, 56–7, 58, 96
diplodocus, 64
dipterus, 56
Doggerland, 68–9, 87
dykes, 20, 32, 44, 45, 46–7, 93, 95
earthquakes, 4, 6, 96, 97
ecological niche, 51, 64, 93
ecosystems, 14, 29, 50, 51–2, 57, 58–9, 61, 67, 86, 93
Edinburgh Castle Rock, 32
Edinburgh New Town, 73
Eigg, Sgurr of, 43, 44
Elginia mirabilis, 62
environmental change, 88–9
erosion, 9, 18, 22, 23, 26, 36, 39, 40, 43, 67, 72, 93–4, 96, 97
extinctions, 51
extrusions, 95
fissures, 31, 41, 94, 95
Flanders Moss, 80
fossils, 3, 48, 50, 53–4, 55–6, 58, 65, 94, 96
freshwaters, 3, 21, 52, 55–6, 95
further reading, 98
gabbro, 34, 36, 40, 76, 94–5
gastropods, 60, 94
Geikie, Sir Archibald, 34–5
geological column, 13, 94, 96
geological legacy, 72–92
geological past, 2–4
geological periods, 14–16
Geological Survey, 16, 27, 28–9, 34, 35, 37, 40, 56
geological time, 4, 9, 16, 50, 94, 95, 96, 97

99

beginnings of, 18–21
immensity of, 6–7, 10, 13
plants and spiders in, 57
geology, 6–7, 9–13, 28–9, 34, 40, 46, 49, 56, 72, 75, 92, 94
glaciers, 68, 70, 72, 74, 76, 89, 94, 95
Glen Ogle, 84
Glencoe, 15, 37, 93
Gleneagles, 74
gley soils, 78, 94
gneiss, 16, 17–22, 28, 34, 35, 95
granites, 7–8, 20, 23, 35–6, 40, 46–7, 70, 72, 73, 94–5
Great Tapestry of Scotland, 17
greenhouse gases, 49, 85–6, 88, 94
habitats, 23, 53, 55, 83, 85, 94, 95, 97
geological processes and, 78–81
Hareheugh Crags, 38
Harker, Dr Alfred, 40
Highland Boundary Fault, 26, 46
hill walking, 74–5
Holmes, Arthur, 11, 12–13
homo sapiens, 14, 51, 67
Hooker, Joseph, 10
Horne, John, 28–9
hornfels, 95
human reach, change and, 85–8
Hutton, James, 6, 7–9, 11, 32, 47
hydrothermal vents, 48, 49
Iapetus Ocean, 15, 16, 17, 24–7, 28, 29, 35–7, 52, 53, 54, 55, 72, 92, 94
ice, 2, 14, 18–19, 22, 24, 32, 34, 36, 39, 45, 47, 52, 67–70, 72, 83
melted ice, 18, 42, 68, 74, 76, 80, 86–7, 89, 94
Ice Age, 14, 16, 52, 67–9, 74, 77, 80, 82, 87, 89–90, 92, 94, 95
idoptilus, 59, 60
igneous rocks, 9, 12, 32, 34–5, 40, 43, 93, 94–5, 97
Industrial Revolution, 72
Inner Hebrides, 2
volcanoes of, 38–9
interglacial period, 14, 89, 94, 95
intrusions, 23, 36, 40, 95
iron, 5
Jurassic period, 15, 42, 63, 64–5, 93
Kelvin, Sir William Thomson, 1st Baron Kelvin, 11–12
Knockan Crag, 28–9

Lairig Ghru, 70
lakes, 2
landmarks, 4
landscapes, 1–2, 7, 14, 18–19, 23–4, 28, 31, 36–7, 44–5, 47, 62, 72, 74, 76–7, 81
ancient landscape, 21–2, 43
dynamism in, 83–5
future prospects, 88–9
Ice Ages, 67–71, 89–92
Jurassic landscape, 63–7
lava types, 33–4
lepidendron, 58
leptopleuron lacertinum, 63
life on Earth, complexity of, 2–3
limestone, 15, 34, 42, 53, 95, 96
'Lizzie the Lizard' *(Westlothiana lizziae)*, 58, 60
Loch Lomond, 73, 83
Lyell, Sir Charles, 9–11
MacCulloch, John, 16, 67
MacCulloch's tree, 66
machair, 81–3
magma, 19, 23, 31–3, 36–8, 40, 42, 43, 44, 45, 46, 47, 95, 96
mantle, 5, 38, 39, 83, 95
meltwater, 74
metamorphic rocks, 19, 23, 75, 95
micro-organisms, 77–8
Miller, Hugh, 56, 64
minerals, 19, 21, 23, 34, 40, 93, 95, 97
moines, 22–5
molluscs, 60, 65, 93, 95
moraine, 70, 95
mountain ranges, 4
muds, 2
Mull, Isle of, 14, 32, 39, 45–6, 66–7, 81
'Munro-bagging,' 74–6
Munro of Lindertis, Sir Hugh, 74–5
Murchison, Professor Roderick Impey, 56
National Museums of Scotland, 57, 58, 61, 65
natural disasters, 6, 89
nature, climate and, 78–81
Neogene period, 14
Newton, Isaac, 7
nickel, 5
Northwest Highlands, 2, 18, 21, 35, 50, 71, 81, 82
Baltica and, final collision, 27–8
obsidian, 33, 94–5
oceans, 2

INDEX

Ordovician period, 15, 26, 51, 52, 53–4, 93, 96
orogeny, 35, 95–6
Outer Hebrides, 2, 16, 18–19, 21, 81
Palaeogene period, 14, 52, 66–7
Peach, Benjamin, 28–9
period, 96
geological periods, 14–16, 92, 93, 96
Permian period, 15, 17, 51, 52, 61, 63, 97
Pettico Wick, 27
Phanerozoic period, 50–51, 94, 96
photosynthesis, 57, 96
pictonia, 65
Planet Earth, 22, 48, 51, 52, 68, 87, 88, 90, 92, 93, 94
plate tectonics, 4, 13, 27, 93, 96
Playfair, John, 8
plesiosaurs, 65
plugs, 37, 47, 96
podzols, 78, 79, 96
power station chimneys, 85
Precambrian period, 13, 57, 94
Proterozoic period, 16
pterichthyodes milleri, 56
pterodactyls, 64
pyroclastic flows, 37, 94–5
Quaternary period, 14, 94
Rest and Be Thankful, 73
rock climbing, 74–5
rock layers, 1, 2
rock types, 9, 76, 94, 95, 96
Rum, Small Islands and, 42–3
sands, 2, 15, 23, 24, 26–7, 29, 40, 43, 52, 53, 55, 62–3, 64, 74, 76, 81, 82–3, 94, 95
rivers of, 21–2
sandstones, 1, 15, 16, 17, 21, 22–3, 42, 50, 52, 55, 56, 62, 75, 95, 96
schists, 16, 34, 95
Scottish Borders, 32, 37, 38, 72, 98
Scottish Natural Heritage, 29, 89
Scott's View (Eildon Hills), 31–2
sea levels, 87
sedimentary rocks, 16, 32, 46, 56, 96, 97
shales, 42, 50, 53, 95, 96
Silurian period, 13, 15, 53, 54, 93, 96
Skara Brae, 17, 71
Skye, 14, 15, 17, 28, 31, 32, 39, 43–4, 64, 65, 75–6, 92
volcanic island of, 40–42
slate, 95
soils, 22, 66–7, 76–8, 79, 94, 96
Solar System, 5
Southern Uplands, 16, 24, 27, 53
sphagnum moss, 78
stagonolepis, 63
strata, 2, 7, 23, 40, 47, 64, 66, 96
stromatolites, 57
subduction, 4, 96
tectonic plates, 4, 5–6, 29, 48, 95, 96
Torridon, 16, 17, 22–3, 24, 50, 75
trace fossil, 96
transportation routes, 72
Traprain Law, 32
Triassic period, 15, 39, 51, 52, 61, 63
trilobites, 3, 50, 53, 54, 93, 97
Trossachs National Park, 83–4
Trotternish, 32, 41–2
tsunamis, 6, 66, 97
tundra, 52, 97
ultrabasic rocks, 40, 97
unconformity, 7, 97
Ussher, James, Archbishop of Armagh, 11
vents, 31–2, 38, 45, 48–9, 96, 97
Victorian scientific inquiry, 1
volcanic ash, 89, 94–5
volcanic bombs, 31, 32, 94–5
volcanoes, 2, 12, 14, 26–7, 30–35, 37–8, 40, 41, 43, 46–7, 49, 66, 94
original volcanoes, 32
volcanic past, landmarks with, 30–31
water column, 58, 97
waters, 2, 3, 16, 27, 47, 48, 52, 54, 57, 58, 95
Western Isles, 18, 20, 81
wind, 2, 9, 15, 19, 22, 32, 39, 47, 62, 72, 78, 81, 82, 87, 88, 93, 94

Picture credits

Title page Martin MO3; ix N Mrtgh; 1 SNH; 2 Martin Rietze/Science Photo Library; 3 (all) NMS; 4 (both) drawn by Robert Nelmes; 5 (both) drawn by Jim Lewis; 7 (left) SNH/Lorne Gill; 8 Alan McKirdy; 13 Our Dynamic Earth; 16 drawn by Jim Lewis; 17 The Great Tapestry of Scotland Charitable Trust; 18 SNH/Angus and Patricia Macdonald; 19 (left) Alfred Pasieka/Science Photo Library; 19 (right) SNH; 20 (both) SNH; 22 (left) SNH/Lorne Gill; 22 (right) N Mrtgh; 23 ABB Photo; 24 (upper) Iain McGillivray; 24 (lower) drawn by Jim Lewis; 25 drawn by Jim Lewis; 27 (top) SNH/Lorne Gill; 28 (top) SNH/Lorne Gill; 29 SNH; 30 cristapper; 31 Science Photo Library; 33 top left jan kranendonk; 33 (top right) Moira McKirdy; 33 (middle left) SNH/Lorne Gill; 33 (middle right) Martin Gillespie; 33 (bottom) Alan McKirdy; 34 (top left) Alfred Pasieka/Science Photo Library; 34 (top right) Mike McNamee/Science Photo Library; 34 (middle) Alfred Pasieka/Science Photo Library; 34 (bottom) Alfred Pasieka/Science Photo Library; 36 SNH/Patricia and Angus Macdonald; 37 John A. Cameron; 38 Alan McKirdy; 39 (both) drawn by Jim Lewis; 42 (bottom) Martin Gillespie; 42 SNH; 43 Martin Gillespie; 44 SNH; 45 (top) SNH/ Patricia and Angus Macdonald; 45 (bottom) SNH; 46 (bottom) SNH; 48 Dr Ken MacDonald/Science Photo Library; 47 Light traveller ; 49 Martin Rietze/Science Photo Library; 53 (all) reproduced with permission from Nigel Trewin, *Scottish Fossils*, Dunedin Academic Press, 2013; 54 (all) reproduced with permission from Nigel Trewin, *Scottish Fossils*, Dunedin Academic Press, 2013; 55 (top) drawn by Robert Nelmes; 55 (bottom) drawn by Jim Lewis; 56 (all left) reproduced with permission from Nigel Trewin, *Scottish Fossils*, Dunedin Academic Press, 2013; 57 (left) drawn by Robert Nelmes; 57 (right, both) reproduced with permission from Nigel Trewin, *Scottish Fossils*, Dunedin Academic Press, 2013; 59 SNH; 60 reproduced with permission from Nigel Trewin, *Scottish Fossils*, Dunedin Academic Press, 2013; 61 drawn by Jim Lewis; 62 (top) Sue Beardmore/Elgin Museum; 62 (bottom) Neil Clark; 63 reproduced with permission from Nigel Trewin, *Scottish Fossils*, Dunedin Academic Press, 2013; 64 (right) Neil Clark; 65 (bottom) reproduced with permission from Nigel Trewin, *Scottish Fossils*, Dunedin Academic Press, 2013; 66 (left) SNH/Lorne Gill; 66 (right) Colin MacFadyen; 68 drawn by Jim Lewis; 69 John Gordon; 70 (both) SNH; 71 Alan McKirdy; 73 (top) Alan McKirdy; 73 (middle) Moira McKirdy; 73 (bottom) Moira McKirdy; 74 Gleneagles; 75 (top) John Gordon; 75 (bottom) Ashley Cooper/Science Photo Library; 76 Moira McKirdy; 77 (both) James Hutton Institute; 79 (all) SNH; 80 SNH; 81 Kevin George; 82 SNH; 83 Bumblebee Conservation Trust